© Rolando Zucchini 2015
© Mnamon 2015
ISBN: 9788869490484

English version edited by Diana Tonini

Cover image: Collatz intensities by Roddy Collins and Andrew Shapira

The pictures have been designed by the author

Rolando Zucchini

The conjecture of Syracuse

MNAMON

Preamble

The conjecture of Syracuse (better known as the Collatz conjecture) is one of the many mathematical conjectures still waiting for proof. In this essay it is addressed by highlighting some of its features. From one of these features takes its cue a process that leads to a theorem whose proof solves the conjecture in a complete and definitive way. With few steps we exit the maze, we reach sea level from high altitudes and we tame the crazy lift of a very high skyscraper. The solution of the conjecture of Syracuse reveals the magical harmony of odd numbers and opens new horizons to the number theory.

Introduction

A conjecture (from the Latin - *coniectura* -, from the verb *conicere*: infer, interpret, conclude) is a statement based on an intuition, a reflection, a flash of genius. The word conjecture (EIKASIA) is found for the first time in the writings of Plato (428-347 BC). After him the Stoics argued that "*il sapiente deve sempre esprimersi per certezze e non per congetture*" ("*the wise man should always speak for certainty and not for speculation*"). To these Cicero (107-44 BC) retorts that "*è proprio del sapiente fare congetture su ciò che ignora*" ("*it is of the wise to speculate on what it ignores*"). Nicholas of Cusa (Kues 1401 - Todi 1464) deals with a systematic relationship between the known and the unknown, and gives value to the incomplete knowledge of conjectures considering them unique and noble. The term conjecture was often used by Karl Popper (Vienna 1902 - London 1994) in the context of scientific philosophy. In mathematics, a conjecture is a proposition considered true but for which there is no proof. Unlike the so-called empirical sciences, mathematics is based on incontrovertible truth, hence the many attempts to prove the most famous conjectures, many of which have been demonstrated in a logical and irrefutable way; many others, however, are still waiting for proof. Among them, the conjecture of Syracuse, despite being tested for the order numbers of $2x10^{12}$, two million of millions.

***

The Syracuse conjecture is mostly known as the Collatz conjecture, named after the mathematician Lothar Collatz (Arnsberg 1910 - Varna 1990) who formulated it in 1937, and since then has not been demonstrated and has never found a counter-example that would make it false. Lothar Collatz studied mathematics at various universities, including the University of Berlin under the rectory of Alfred Klose (1895-1953). He graduated in 1935 with a thesis on approximate solutions of linear differential equations. He was an honorary member of the Mathematical Society of Hamburg and awarded an honorary degree by the University of Sao Paulo, Brazil, from that of Dundee (Scotland), and from the Technical Universities of Vienna, Hannover and Dresden. Around the thirties of the twentieth century he was responsible for the theory of numbers, and it was during these studies that he formulated the conjecture. His friend Helmut Hasse (Kassel 1989 - Ahrensburg 1979), in the 50s, presented the issue at a meeting of mathematicians at the University of Syracuse (New York), hence the name *conjecture of Syracuse*. It is also known by the name of the *problem of Ulam*, from mathematician Stanislaw Ulam (1909 - 1984) who proposed it to the students of the University of Los Alamos (New Mexico) where he taught. The problem of Ulam was picked up, in the 60s, by the Japanese mathematician Shizuo Kakutani (1911 - 2004), and for this it is also known as the *problem of Kakutani*. The conjecture of Syracuse is mentioned in the film "*La donna che canta*" (2010) by Denis Villeneuve, adapted from the play "*Incendies*" by Wajdi Mouawad.

The conjecture of Syracuse

The conjecture of Syracuse states: *If at any natural integer n, non- zero, we will apply the algorithm 3n+1 if n is odd, n/2 if n is even, the sequence of the values obtained, precipitates to 1 after a finite number of steps, always in compliance with the final cycle {4; 2; 1}.*

$$\forall n \in \mathbb{N} : n \neq 0 \longrightarrow \begin{cases} 3n+1 & \text{se } n \text{ è dispari} \\ \\ n/2 & \text{se } n \text{ è pari} \end{cases}$$

Chosen n = 12 (even), applying the algorithm of Collatz we obtain the following sequence:

n = 12 → S(12) = {12; 6; 3; 10; 5; 16; 8; 4; 2; 1}

the number 12 falls down to 1 in nine steps.

The sequence (or succession) generated by the number 12 is oscillating (neither increasing nor decreasing) and can be represented with a graph on a Cartesian plane having its origin (0,1). The number 1 can be understood as the minimum level or the sea level. The steps are shown on the horizontal axis and the generating number on the vertical axis. The graph of fluctuation of the sequence S(12) (generated by the number 12) is represented in figure 1. In it the horizontal line parallel to the abscissa axis (distance) is the horizon of the number 12. The number 12, during the oscillation of the sequence S(12) by it generated, exceeds its horizon once, exactly in step 5, in correspondence of which reaches a peak 16.

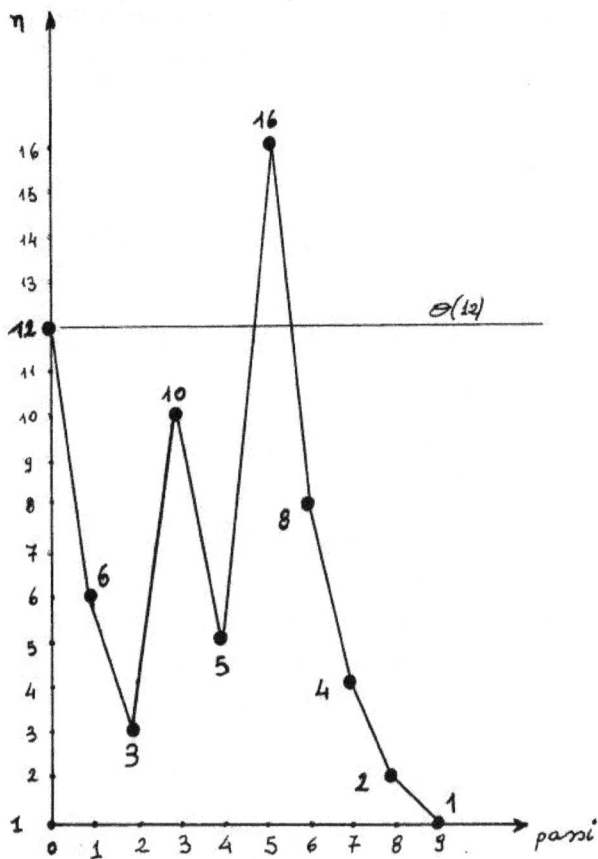

Fig. 1

Chosen n = 7 (odd), applying the algorithm we obtain the sequence:

n = 7 → S(7) = {7; 22; 11; 34; 17; 52; 26; 13; 40; 20; 10; 5; 16; 8; 4; 2; 1}

the number 7 falls to 1 in 16 steps.

Figure 2 shows the graph regarding the oscillation of the sequence S(7).

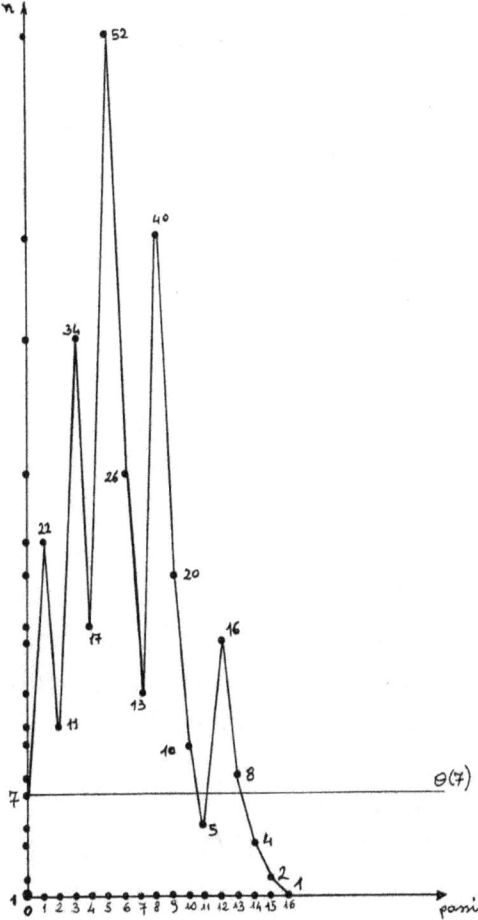

Fig. 2

It is to be noted that if you choose as starting number an odd number, the first element of the sequence is always an even number, since the product of 3 for another odd number still gives an odd number, which becomes even with the addition of 1. This happens for all the successors of an odd number contained in the sequence. An even number, however, may be followed by an odd number or by one or more even numbers.

$$n = 48 \rightarrow S(48) = \{48; 24; 12; 6; 3; 10; 5; 16; 8; 4; 2; 1\}$$

If you choose an even number power of 2, $n = 2^p = 2$; 4; 8; 16; 32; 64; 128; 256; 512; 1024; 2048; ..., it reaches 1 after a cycle of p applications of the algorithm. For example:

$2^4 = 16 \rightarrow S(16) = \{16; 8; 4; 2; 1\}$ (5 values, 4 steps)
$2^7 = 128 \rightarrow S(128) = \{128; 64; 32; 16; 8; 4; 2; 1\}$ (8 values, 7 steps)

The same observation also applies to the odd numbers, when $3n+1 = 2^p$, i.e. $n = (2^p-1)/3$. In this case the number of steps needed to get it to fall to 1 is equal to p+1. Bearing in mind that the algorithm is applicable only to natural integers, let's see some examples:

$p = 1 \rightarrow n$ is not an integer
$p = 2 \rightarrow n = 1 \rightarrow \{1; 4; 2; 1\}$ (4 values, 3 steps)
$p = 3 \rightarrow n$ is not an integer

$p = 4 \rightarrow n = 5 \rightarrow S(5) = \{5; 16; 8; 4; 2; 1\}$ (6 values, 5 steps)

$p = 5 \rightarrow$ n is not an integer

$p = 6 \rightarrow n = 21 \rightarrow S(21) = \{21; 64; 32; 16; 8; 4; 2; 1\}$ (8 values, 7 steps)

$p = 7 \rightarrow$ n is not an integer

$p = 8 \rightarrow$ n is not an integer

$p = 9 \rightarrow$ n is not an integer

$p = 10 \rightarrow n = 341 \rightarrow S(341) = \{341; 1024; 512; 256; 128; 64; 32; 16; 8; 4; 2; 1\}$ (12 values, 11 steps)

...

Here we stop, but it is interesting to note that for some natural numbers, either odd or even, which constitute a partition of N (v. Fig. 3), it is possible to establish *a priori* the number of values contained in their respective sequences obtained by applying to them the algorithm of Collatz, and then establish the exact number of steps needed to reach 1.

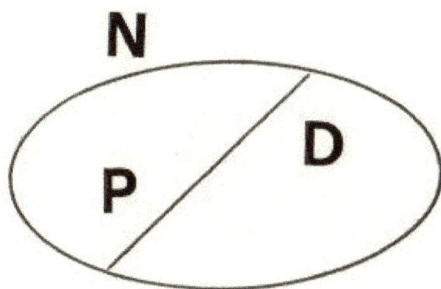

Fig. 3

Precisely, indicating the set of even numbers P and D the set of odd numbers:

If $n \in P \subset N : n = 2^p \Rightarrow$ the corresponding sequence $S(n)$ contains p+1 values and it reaches 1 in p steps.

If $n \in D \subset N : n = (2^p-1)/3 \Rightarrow$ the corresponding sequence $S(n)$ contains p+2 values and it reaches 1 in p+1 steps.

Things get complicated if the choice is a number that does not satisfy the above conditions. For it is not possible to establish the number of steps needed to get to 1. If we choose, for example, the number 25 we will have the following sequence:

25 → 76 → 38 → **19** → 58 → 29 → **88** → 44 → 22 → 11 → 34 → **17** → 52 → 26 → 13 → 40 → 20 → 10 → 5 → 16 → 8 → **4** → **2** → **1**

It consists of 23 steps. We observe that the sequence of values does not contain two same numbers, but they are all different from each other. This happens for all sequences of all the numbers. In short, the sequences obtained through the algorithm of Collatz precipitate to 1 taking different values between them. This is obvious. In fact, if in a succession two numbers were the same, from them onwards we would have the same sequence, giving rise to a vicious circle, and this is impossible (clearly with the exclusion of the number 1 that generates the final cycle). It should be noted also that, when a succession is built, many more are known. With reference to the previous example, the sequence of the number 25 allows us to know the sequences of numbers:

19 → 58 → 29 → **88** → 44 → 22 → 11 → 34 → **17** →
52 → 26 → 13 → 40 → 20 → 10 → 5 → 16 → 8 → **4**
→ **2** → **1**
88 → 44 → 22 → 11 → 34 → **17** → 52 → 26 → 13 →
40 → 20 → 10 → 5 → 16 → 8 → **4** → **2** → **1**
17 → 52 → 26 → 13 → 40 → 20 → 10 → 5 → 16 →
8 → **4** → **2** → **1**

And so on for all the others. The sequence of number 25 then allows us to know the sequences of the other twenty natural numbers odd or even (excluding the values of final cycle: 4, 2, 1 and 25). If a number generates a sequence of 1000 values, from it will know the sequences of the 996 numbers in it.

When calculating the values of the sequences we can also observe that each of them connects to a previous one, when one of its values becomes equal to that of another already computed. Figure 4 shows the diagram of such a consideration in reference to the consecutive numbers 5, 6, 7.

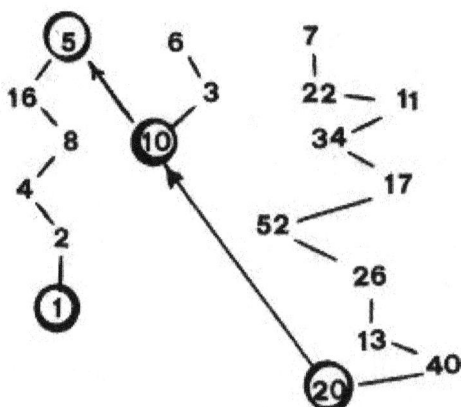

Fig. 4

The sequence of number 7, from the value 20, continues in that of number 6, which in turn, from value 10, continues in that of 5. If we would continue the computation of successive sequences we would realize that the sequence of number 8 is already present in the sequence of number 5, while that of number 9, from the value 14, continues in that of number 7, which in turn ... ... and so on.

Again with reference to the number 25, however this reasoning is valid for any other number, another important observation we can make is that when the value of the sequence falls below 25, in this case 19, its succession from here on becomes perfectly equal to that of this number. In this way the sequence of the number 25 is linked to the succession of the number 19. In short, the endless sequences generated by the algorithm give rise to a kind of interlacement, in which each of the sequences is linked to one of the previous ones. In Figure 5 its shown the interlacement produced by the first seven natural numbers.

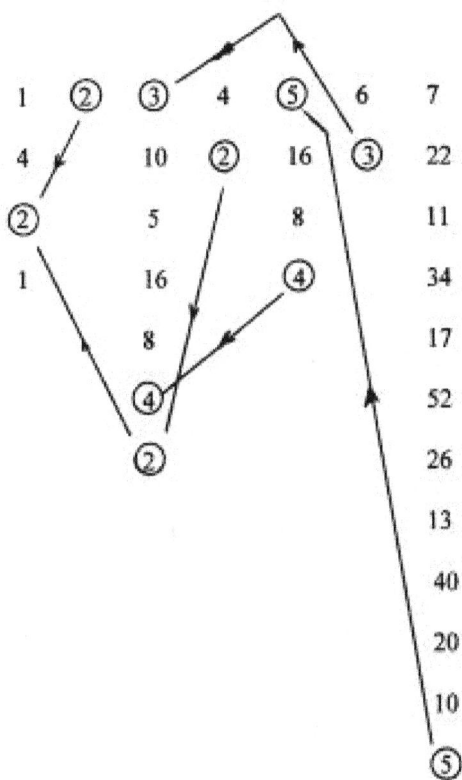

Fig. 5

The numbers 2 and 3 snap to the sequence of number 1, those of number 4, number 5, and number 6 to that of number 3, while the sequence generated by number 7 attaches to the one generated by number 5. We could build an intertwining of hundreds of consecutive numbers.

***

If we take 33 as generating number, applying the algorithm, we obtain the sequence:

33 $\rightarrow$ 100 $\rightarrow$ 50 $\rightarrow$ **25** < 33
25 $\rightarrow$ 76 $\rightarrow$ 38 $\rightarrow$ **19** < 25
19 $\rightarrow$ 58 $\rightarrow$ 29 $\rightarrow$ 88 $\rightarrow$ 44 $\rightarrow$ 22 $\rightarrow$ **11** < 19
11 $\rightarrow$ 34 $\rightarrow$ 17 $\rightarrow$ 52 $\rightarrow$ 26 $\rightarrow$ 13 $\rightarrow$ 40 $\rightarrow$ 20 $\rightarrow$ **10** < 11
10 $\rightarrow$ **5** < 10
5 $\rightarrow$ 16 $\rightarrow$ 8 $\rightarrow$ **4** < 5
4 $\rightarrow$ **2** $\rightarrow$ **1** final cycle. (v. Fig. 6).

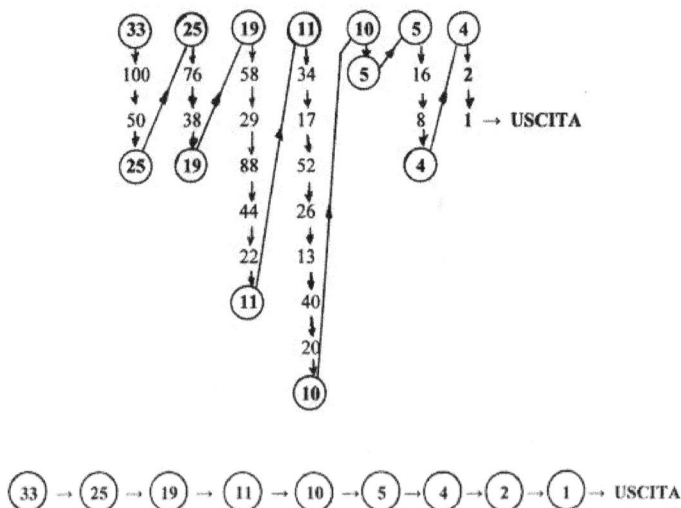

Fig. 6

If we take 34 as generating number, applying the algorithm, we obtain the sequence:

34 → **17** < 34
17 → 52 → 26 → **13** < 17
13 → 40 → 20 → **10** < 13
10 → **5** < 10
5 → 16 → 8 → **4** < 5
4 → **2** → **1** final cycle. (v. Fig. 7)

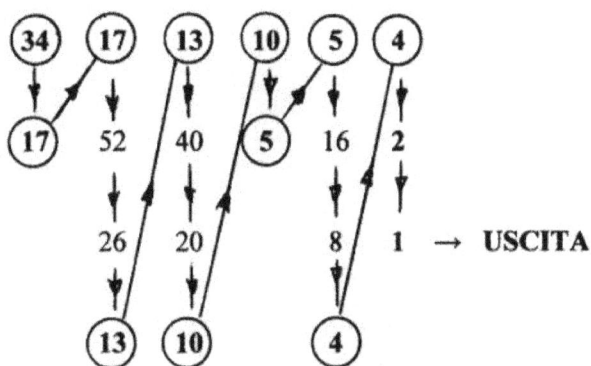

Fig. 7

If we take 49 as generating number, applying the algorithm, we obtain the sequence:

49 → 148 → 74 → **37** < 49
37 → 112 → 56 → **28** < 37
28 → **14** < 28
14 → **7** < 14

$7 \rightarrow 22 \rightarrow 11 \rightarrow 34 \rightarrow 17 \rightarrow 52 \rightarrow 26 \rightarrow 13 \rightarrow 40 \rightarrow$
$20 \rightarrow 10 \rightarrow \mathbf{5} < 7$
$\mathbf{5} \rightarrow 16 \rightarrow 8 \rightarrow \mathbf{4} < 5$
$\mathbf{4} \rightarrow 2 \rightarrow 1$ final cycle. (v. Fig. 8)

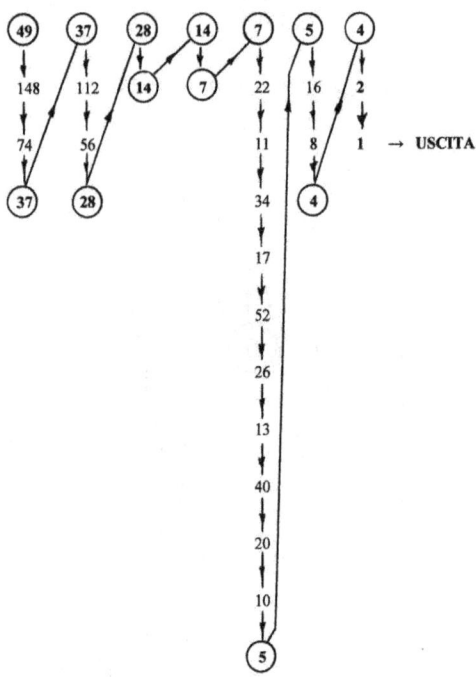

Fig. 8

If we take 204 as generating number, applying the algorithm, we obtain the sequence:
$\mathbf{204} \rightarrow \mathbf{102} < 204$
$\mathbf{102} \rightarrow \mathbf{51} < 102$

$51 \rightarrow 154 \rightarrow 77 \rightarrow 232 \rightarrow 116 \rightarrow 58 \rightarrow \mathbf{29} < 51$
$\mathbf{29} \rightarrow 88 \rightarrow 44 \rightarrow \mathbf{22} < 29$
$\mathbf{22} \rightarrow \mathbf{11} < 22$
$\mathbf{11} \rightarrow 34 \rightarrow 17 \rightarrow 52 \rightarrow 26 \rightarrow 13 \rightarrow 40 \rightarrow 20 \rightarrow \mathbf{10} < 11$
$\mathbf{10} \rightarrow \mathbf{5} < 10$
$\mathbf{5} \rightarrow 16 \rightarrow 8 \rightarrow \mathbf{4} < 5$
$\mathbf{4} \rightarrow \mathbf{2} \rightarrow \mathbf{1}$ final cycle. (v. Fig. 9

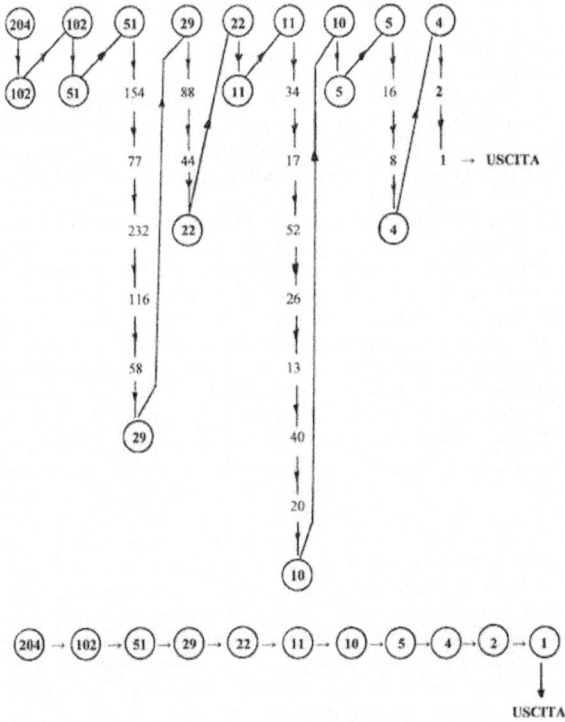

Fig. 9

Such a proceeding is applicable to $\forall$ n $\in$ N: n $\neq$ 0 and n $\neq$ 1.

In brief, the endless sequences can be imagined as a sort of maze in which the various numbered rooms are connected to each other in a descending order and according to the rule set by the algorithm, up to the final cycle **4** → **2** → **1** and finding the exit.

The conjecture of Syracuse, then would be proven if we could demonstrate that any sequence contains a lesser value than the starting number (generating number). Because if this happens it hooks to this, which in turn hooks to another lower than it, and so on, up to the point where we precipitate necessarily to 1 (exit of the maze). Or, as they say, the main horizon of the generating number descends to the next lower horizons until we reach the sea level. This can be stated with certainty for even numbers equal to $n \in P \subset N$: $n = 2^p$ and for odd numbers $n \in D \subset N$: $n = (2^p-1)/3$. For all other numbers there is uncertainty. However, the consideration previously explained is interesting and worthy of further in-depth analysis.

In depth-analysis

1)
A numerical sequence is an infinite set of values generated by a mathematical law

$$f: n \rightarrow f(n)$$

and indicated by $S = a_0; a_1; a_2; \ldots ; a_n; a_{n+1}; \ldots$.
Generally the function f has by domain the set of natural numbers N (with or without the zero), while the codomain may be the same N or another numerical set. The function $f: n \rightarrow n/2$ has as codomain the set of rational numbers Q:

$$S = 0; 1/2; 1; 3/2; \ldots ; n/2; (n+1)/2; \ldots$$

The function $f: n \rightarrow \sqrt{n}$ has as codomain the set of real numbers R:

$$S = 0; 1; \sqrt{2}; \sqrt{3}; 2; \sqrt{5}; \ldots ; \sqrt{n}; \sqrt{(n+1)}; \ldots$$

To determine a sequence, thus, you must specify an analytical expression of the type $a_n = f(n)$ which, after a finite number of mathematical operations, allows you to calculate the term a_n order of sequence starting from the value of n. If we consider the sequence generated by the algorithm:
1) $a_1 = 1$ 2) $a_{n+1} = (a_n+1)/a_n$
Its terms are:

$a_1 = 1$
$a_2 = (1+1)/1 = 2$
$a_3 = (2+1)/2 = 3/2$

$a_4 = (3/2+1)/3/2 = 5/3$

$a_5 = (5/3+1)/5/3 = 8/5$

...

and so on.

If the terms of a succession grow with the increase of the index, that is, if $i < j \Rightarrow a_i < a_j$ then the sequence is increasing. If, on the other hand, the terms of a succession decrease with the increase of the index, that is, if $i < j \Rightarrow a_i > a_j$ then the sequence is decreasing.

To determine whether a sequence is increasing we can just check that the generic term is less than its successor, i.e.: $a_n < a_{n+1}$. To determine if a sequence is decreasing we can just check that the generic term is greater than its successor, i.e.: $a_n > a_{n+1}$.

Increasing or decreasing sequences are called monotonic. A succession that is not monotonic it is called oscillating.

The sequence $a_n = n/(n+1)$ it is an increasing monotonic. Its elements are: 0; 1/2; 2/3; 3/4; 4/5; ... ; $n/(n+1)$; $(n+1)/(n+2)$; ... and is $a_n < a_{n+1}$. Indeed:

$n/(n+1) < (n+1)/(n+2) \rightarrow (n+1)/(n+2) - n/(n+1) > 0 \rightarrow (n+1)^2 - n(n+2) > 0 \rightarrow n^2+2n+1-n^2-2n > 0 \rightarrow 1 > 0$ true. So $a_n < a_{n+1}$ it's always true.

Some sequences become monotonic from a certain stage forward. For example the sequence generated by the analytical expression $a_n = n^2-6n$ has as values:

0; -5; -8; -9; -8; -5; 0; 7; 16; 27; ...

which by the end $a_3 = -9$ turns out to be an increasing monotonic. In fact, for being $n \geq 3 \rightarrow a_n < a_{n+1} \rightarrow n^2-6n$

$< (n+1)^2-6(n+1) \rightarrow n^2-6n < n^2+2n+1-6n-6 \rightarrow 2n-5 > 0$
which being n ≥ 3 it's always true.

2)
The principle of induction is an important demonstration technique frequently used in mathematics. It allows us to demonstrate that a proposition p is true if $p(1)$ is true, and presumed true $p(n)$, proves to be true $p(n+1)$. I.e.:

If:
1) $p(1)$ is true
2) $p(n)$ is true by hypothesis.
3) $p(n+1)$ is true for demonstration.
Then: the proposition $p(n)$ is true $\forall$ n $\in$ N.
Let's clarify the induction principle with an example.
Demonstrate that the proposition p: the sum of the first n natural numbers (excluding zero) is:
$s(n) = n(n+1)/2$.
$p(1)$ is true. In fact $s(1) = 1 \cdot (1+1)/2 = 1 \cdot 2/2 = 1$
$p(2)$ is true. In fact $s(2) = 1+2 = 2 \cdot (2+1)/2 = 6/2 = 3$.
$p(3)$ is true. In fact $s(3) = 1+2+3 = 3 \cdot (3+1)/2 = 12/2 = 6$
...
But proceeding like this would take us an endless amount of checks of the truth fullness of the proposition p. Let's apply the principle of induction.
1) $p(1)$ is true
2) $p(n)$ is supposed true
3) we demonstrate that it is true $p(n+1)$ which is that $s(n+1) = 1+2+3+4+ \ldots +n+(n+1) = (n+1)(n+2)/2$.

Having supposed true s(n) = 1+2+3+4+ ... +n = n(n+1)/2 adding to both members of equality (n+1) we have: s(n+1) = 1+2+3+4+ ... +n+(n+1)= n(n+1)/2 + (n+1) = [n(n+1)+2(n+1)]/2 = (n+1)(n+2)/2

Resulting p(n+1) real, then the proposition p(n) is true ∀ n ∈ N. QED

The principle of induction is clear but not demonstrable. It is an axiom, the fifth postulate, formulated by Giuseppe Peano (Spinetta (Cuneo) 1858 - Torino 1932) in his theory on the axiomatization of arithmetic.

Sometimes the property (or proposition) that you want to prove only applies to natural integers greater than some n_0. In this case the induction principle can be formulated taking as domain N* = {∀ n ∈ N: n > n_0}.

3)

A recursive method is a process that refers to itself but does not result in a vicious circle. In order to be profitable it is necessary that the chain of self-references arrives to an end. A succession is said recursive if: a_1 = f(a_0); a_2 = f(a_1); ...; a_n = f(a_{n-1}); ... .

An iterative procedure is a procedure based on the repetition of one or more mathematical operations to a finite number of times. A succession is said iterative when attached to a_0 = k, then: a_1 = f(k) or a_1 = g(k) or ...; a_2 = f(a_1) or a_2 = g(a_2) or ...; a_n = f(a_{n-1}) or a_n = g(a_{n-1}), ... ; ... . The first element k is the generator of the succession. The sequences obtained by the algorithm of Collatz are iterative.

4)

If we state by p an even number and d an odd number, the following rules apply:

$p_1 + p_2 = p_3$
$p + d = d$
$d_1 + d_2 = p$
$p_1 \cdot p_2 = p_3$
$p \cdot d = p$
$d_1 \cdot d_2 = d_3$

Demonstration of the conjecture of Syracuse

We have seen that the sequences generated by the algorithm of Collatz are oscillating, excluding those whose generating number is power of 2. The latter are decreasing monotonic and fall down to 1 after a number of steps previously calculated. For the oscillating successions we found that they certainly fall down to 1 if a term of the sequence, generated by a generic number n, assumes a smaller value of n, so below the horizon of n that we can mark with $o(n)$.

Marking with n_i that value ($n_i < n$; $o(n_i)$ lower of $o(n)$), the sequence generated by n is linked to that generated by n_i, which, in turn, is linked to that of $n_j < n_i$ ($o(n_j)$ lower of $o(n_i)$), which ... ., and so it precipitates surely to 1, exiting from the labyrinth, or, as it's said, reaches the level of the sea. We wonder: does this happens in all sequences? In any succession there is always an element that acquires a smaller value of the generator, that is of the number that has generated it? This is certainly true for the even numbers. For them, the first element of the sequence is n/2, and n/2 < n. If, instead, the genereting number is odd then the first element of the sequence is 3n+1, which, by virtue of the algorithm, is certainly even and therefore divisible by 2. That is: $3n+1 \rightarrow (3n+1)/2$. Unless it is not of the type $(3n+1)/2 = 2^p$ or $(3n+1)/2 = (2^p-1)/3$ ($n \in D \subset N$), it turns out to be $(3n+1)/2 > n$, as is easy to verify. But if 3n+1 was doubly even (twice divisible by two, or, as it were, divisible by 4), triply even (three times divisible by 2, or, as it were, divisible by 8), and then $(3n+1)/4 < n$ and $(3n+1)/8 < n$ they would definitely be true.

Theorem 2n+3

Let's apply the principle of induction the following proposition p:

Any sequence generated by the algorithm of Collatz always contains a term less in value than its generator.

As I said before the proposition p is true for the natural numbers $P \subset N$.

Let's then take into consideration only the odd generating numbers with the exclusion of 1, which generates the final cycle {4; 2; 1} and so it's trivial. We denote the generic odd number with $2n+1$ and its successor with $2n+3$. The principle of induction translates into the following steps:

1) $p(3)$ it is true. In fact: $S(3)$: $3 \rightarrow 10 \rightarrow 5 \rightarrow 16 \rightarrow 8 \rightarrow 4 \rightarrow 2 < 3$

2) let's presume real $p(2n+1)$, so $\exists\ a_n \in S(2n+1)$: $a_n < 2n+1$

3) let's prove that $p(2n+3)$ is true.

Bearing in mind that in the sequences generated by the algorithm of Collatz any odd number has as successor an even number, we will have.

$p(2n+3)$: $2n+3 \rightarrow 3(2n+3)+1 = 6n+10 \rightarrow 3n+5$.

$3n+5$ can be even or odd.

a) If $3n+5$ is even, then $3n+5 \rightarrow (3n+5)/2 < 2n+1$ ($3n+5<$

4n+2 from which n − 3 > 0 it is true for n > 3), then $p(2n+3)$ is true being true $p(2n+1)$. So: $(3n+5)/2 < 2n+1 < 2n+3 \rightarrow (3n+5)/2 < 2n+3$

But 3n+5 is even if n $\in$ D = {1; 3; 5; 7; 9; 11, 13; 15; ... ; 2n-1; 2n+1; ...}.

It follows that p(2n+3) is true in D_1 = {**5; 9; 13; 17; 21; 25; 29; 33; ... ; 4n+1; 4n+5; ...**}.

b) If 3n+5 is odd, then: $3n+5 \rightarrow 3(3n+5)+1 = 9n+16$, which being even in virtue of the algorithm, $9n+16 \rightarrow (9n+16)/2$, which may be even or odd.

b1) If $(9n+16)/2$ is even, then $(9n+16)/2 \rightarrow (9n+16)/4 < 4n+2 = 2(2n+1)$. If this happens $p(2n+3)$ is true, being true $p(2n+1)$. And this happens if $(9n+16)/2$ is at least doubly even (divisible by 4). So: $(9n+16)/2 \rightarrow (9n+16)/4 \rightarrow (9n+16)/8 < 2n+1 < 2n+3 \rightarrow (9n+16)/8 < 2n+3$.

But $(9n+16)/2$ is even for n $\in$ P_1 = {4; 8, 12, 16, 20, 24; 28; 32; ...; 4n; 4n+4; ...}, then in its iterative cycle there is always an element at least divisible by 4.

It follows that $p(2n+3)$ is true in: D_2 = {**11; 19; 27; 35; 43; 51; 59; 67; ...; 8n+3; 8n+11; ...**}.

b2) If $(9n+16)/2$ is odd, then $(9n+16)/2 \rightarrow 3((9n+16)/2)+1 = (27n+48)/2+1 = (27n+50)/2$, which being even in virtue of algorithm: $(27n+50)/2 \rightarrow (27n+50)/4 < 8n+4 = 4(2n+1)$. If this happens $p(2n+3)$ is true being true $p(2n+1)$. And this happens if $(27n+50)/2$ is at least triply even (divisible by 8). So: $(27n+50)/2 \rightarrow (27n+50)/4 \rightarrow (27n+50)/8$

$\rightarrow (27n+50)/16 < 2n+1 < 2n+3 \rightarrow (27n+50)/16 < 2n+3$.
But $(27n+50)/2$ is even for $n \in P_2$ = {2; 6; 10; 14; 18; 22; 26; 30; …; 4n-2; 4n+2; …}, then in its iterative cycle there is always a term at least divisible by 8.
But $(9n+16)/2$ is odd for $n \in P_2$ = {2; 6; 10; 14; 18; 22; 26; 30; …; 4n-2; 4n+2; …}. It follows that $p(2n+3)$ is true in D_3 = {7; 15; 23; 31; 39; 47; 55; 63; …; 8n-1; 8n+7; …}.

We have identified three groups of odd numbers in which $p(2n+3)$ is true. They are:

D_1 = {5; 9; 13; 17; 21; 25; 29; 33; … ; 4n+1; 4n+5; …}
D_2 = {11; 19; 27; 35; 43; 51; 59; 67; …; 8n+3; 8n+11; …}
D_3 = {7; 15; 23; 31; 39; 47; 55; 63; …; 8n-1; 8n+7; …}

But $D_1 \cup D_2 \cup D_3$ = D^* = {5; 7; 9; 11; 13; 15; 17; 19; 21; 23; 25; 27; 29; 31; …; 2n+1; 2n+3; …} = D - {1; 3}, with 1 as generator of the end cycle and so trivial and $p(3)$ verified as true in point 1) of the principle of induction. We can therefore conclude that $p(2n+3)$ is true for all odd numbers D.

Having seen that the proposition $p(n)$ is true for the even numbers P, having shown that it is true for the odd numbers D, being $P \cup D = N$ (see Fig. 10) then $p(n)$ is true $\forall$ $n \in N$. $\rightarrow$ QED

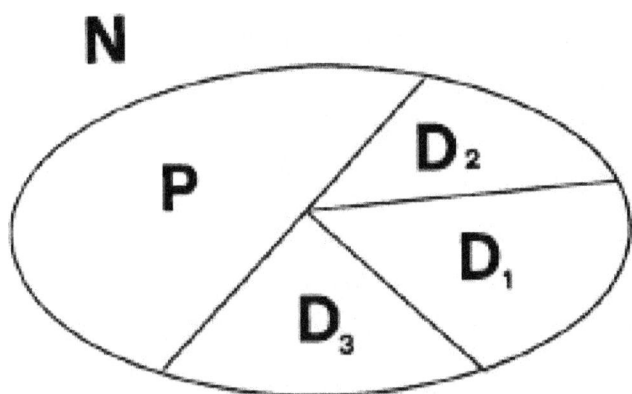

Fig. 10

Theorem 2n+1

In the previous proof that solves the conjecture of Syracuse it has been applied to the induction principle, but it can be avoided by taking into account the generic odd number 2n+1 and proceeding in a perfectly analogous.

Statement::
Any sequence generated by the algorithm of Collatz applied to a generic odd number 2n+1 always contains an a_n term less than its generator.

Demonstration::
Bearing in mind that in the sequences generated by the algorithm of Collatz any odd number has as its successor an even number, we will have:

$$2n+1 \rightarrow 3(2n+1)+1 = 6n+4 \rightarrow 3n+2.$$

3n+2 can be even or odd.

a) If 3n+2 is even, then $3n+2 \rightarrow (3n+2)/2 < 2n+1$ ($3n+2 < 4n+2$, from which $n > 0$, always true).
But 3n+2 is even if $n \in P = \{2; 4; 6; 8; 10; 12; 14; 16; \dots ; 2n; 2n+2; \dots \}$.
It follows that 2n+1 is in $D_1 = \{\mathbf{5; 9; 13; 17; 21; 25; 29; 33; \dots ; 4n+1; 4n+5; \dots}\}$.

b) If 3n+2 is odd, then: $3n+2 \rightarrow 3(3n+2)+1 = 9n+7$, which being even in virtue of algorithm, $9n+7 \to (9n+7)/2$, which may be even or odd.

b1) If $(9n+7)/2$ is even, then $(9n+7)/2 \to (9n+7)/4 < 4n+2 = 2(2n+1) \to (9n+7)/8 < 2n+1$. And this happens if $(9n+7)/2$ is at least doubly even (divisible by 4). So: $(9n+7)/2 \to (9n+7)/4 \to (9n+7)/8 < 2n+1$.
But $(9n+7)/2$ is even for $n \in D_1 = \{1; 5; 9; 13; 17; 21; 25; 29; \ldots; 4n+1; 4n+5; \ldots\}$, then in its iterative cycle there is always an element at least divisible by 4.
It follows that $2n+1$ is in $D_2 = \{3; 11; 19; 27; 35; 43; 51; 59; \ldots; 8n+3; 8n+11; \ldots\}$.

b2) If $(9n+7)/2$ is odd, then $(9n+7)/2 \to 3((9n+7)/2)+1 = (27n+21)/2+1 = (27n+23)/2$, which being even in virtue of the algorithm: $(27n+23)/2 \to (27n+23)/4 < 8n+4 = 4(2n+1) \to (27n+23)/16 < 2n+1$. And this happens if $(27n+23)/2$ is at least triply even (divisible by 8). So $(27n+23)/2 \to (27n+23)/4 \to (27n+23)/8 \to (27n+23)/16 < 2n+1$.
But $(27n+23)/2$ is even for $n \in D_4 = \{3; 7; 11, 15; 19; 23; 27; 31; \ldots ; 4n-1; 4n+3; \ldots\}$, then in its iterative cycle there is always a term at least divisible by 8.
But $(9n+7)/2$ is odd for $n \in D_4 = \{3; 7; 11, 15; 19; 23; 27; 31; \ldots ; 4n-1; 4n+3; \ldots\}$.
It follows that $2n+1$ is in $D_3 = \{7; 15; 23; 31; 39; 47; 55; 63; \ldots; 8n-1; 8n+7; \ldots\}$.

We have identified three groups of odd numbers in which

33

iterative cycle there is always a term $a_n < 2n+1$. They are:

$D_1 = \{5; 9; 13; 17; 21; 25; 29; 33; \dots ; 4n+1; 4n+5; \dots\}$
$D_2 = \{3; 11; 19; 27; 35; 43; 51; 59; \dots; 8n+3; 8n+11; \dots\}$
$D_3 = \{7; 15; 23; 31; 39; 47; 55; 63; \dots; 8n-1; 8n+7; \dots\}$

But $D_1 \cup D_2 \cup D_3 = D^* = \{3; 5; 7; 9; 11; 13; 15; 17; 19; 21; 23; 25; 27; 29; 31; \dots; 2n+1; 2n+3; \dots\} = D - \{1\}$, with 1 as generator of the end cycle and so trivial.
We can therefore conclude that $\exists\, a_n \in S(2n+1): a_n < 2n+1$, $\forall\, (2n+1) \in D \rightarrow$ QED

Explanatory notes

a)

The proofs of the theorem 2n+1 and of the theorem 2n+3, $\forall$ n $\in$ N: n $\neq$ 0, require the condition that in iterative cycles of odd numbers, obtained by applying the algorithm of Collatz, there must be terms either at least doubly even or at least triply even, that become smaller than the generating number, so: $\exists$ a_n $\in$ S(2n+1): a_n < 2n+1, $\exists$ a_n $\in$ S(2n+3): a_n < 2n+3, but this, as shown (sooner or later happens even in very long cycles), it's always true. Formulas that attest this, obtained from the demonstrations, are listed and clarified below.

Theorem 2n+1

2n+1 $\in$ D_1 = {5; 9; 13; 17; ...; 4n+1; ...}, 3n+2 is even if n $\in$ P = {2; 4; 6; 8; ...; 2n; ...}.

Examples:

n = 2 $\rightarrow$ 2n+1 = 5 and **3n+2** = 8 $\rightarrow$ **4** < 5

n = 4 $\rightarrow$ 2n+1 = 9 and **3n+2** = 14 $\rightarrow$ **7** < 9

n = 6 $\rightarrow$ 2n+1 = 13 and **3n+2** = 20 $\rightarrow$ **10** < 13

n = 8 $\rightarrow$ 2n+1 = 17 and **3n+2** = 26 $\rightarrow$ **13** < 17

... ...

n = **2n** $\rightarrow$ 2n+1 = **4n+1** and 3n+2 = **3(2n)+2** = 6n+2 $\rightarrow$ **3n+1** < 4n+1

... ...

$2n+1 \in D_2 = \{3; 11; 19; 27; \ldots; \mathbf{8n+3}; \ldots\}$, $\mathbf{(9n+7)/2}$ it is at least doubly even if $n \in D_1 = \{1; 5; 9; 13; \ldots; \mathbf{4n+1};$ $\ldots\}$. This happens if $(9(\mathbf{4n+1})+7)/2 = 18n+8 \rightarrow 9n+4$ is even, and this happens if $n \in P = \{2; 4; 6; 8; \ldots; \mathbf{2n}; \ldots\}$.

Examples:

$n = 2 \rightarrow 8n+3 = \mathbf{19}$ and $\mathbf{4n+1} = 9 \rightarrow \mathbf{(9n+7)/2} = 44 \rightarrow$ $22 \rightarrow \mathbf{11} < 19$

$n = 4 \rightarrow 8n+3 = \mathbf{35}$ and $\mathbf{4n+1} = 17 \rightarrow \mathbf{(9n+7)/2} = 80 \rightarrow$ $40 \rightarrow \mathbf{20} < 35$

$n = 6 \rightarrow 8n+3 = \mathbf{51}$ and $\mathbf{4n+1} = 25 \rightarrow \mathbf{(9n+7)/2} = 116 \rightarrow$ $58 \rightarrow \mathbf{29} < 51$

$n = 8 \rightarrow 8n+3 = \mathbf{67}$ and $\mathbf{4n+1} = 33 \rightarrow \mathbf{(9n+7)/2} = 152 \rightarrow$ $76 \rightarrow \mathbf{38} < 67$

… …

$n = \mathbf{2n} \rightarrow 8n+3 = \mathbf{16n+3}$ and $\mathbf{4n+1} = \mathbf{8n+1} \rightarrow$ $(9(8n+1)+7)/2 = 36n+8 \rightarrow 18n+4 \rightarrow \mathbf{9n+2} < 16n+3$

… …

Further analysis on the cycles of the connections of odd numbers in D_2 are carried out in note b) and note c2).

$2n+1 \in D_3 = \{7; 15; 23; 31; \ldots; \mathbf{8n-1}; \ldots\}$, $\mathbf{(27n+23)/2}$ it is at least triply even if $n \in D_4 = \{3; 7; 11; 15; \ldots; \mathbf{4n-1};$ $\ldots\}$. This happens if $(27(\mathbf{4n-1})+23)/2 = 54n-2 \rightarrow 27n-1$ is doubly even, and this happens if $n \in D_4 = \{3; 7; 11; 15;$ $\ldots; \mathbf{4n-1}; \ldots\}$.

Examples:

n = 3 → 8n-1 = **23** and **4n-1** = 11 → **(27n+23)/2** = 160 → 80 → 40 → **20** < 23

n = 7 → 8n-1 = **55** and **4n-1** = 27 → **(27n+23)/2** = 376 → 188 → 94 → **47** < 55

n = 11 → 8n-1 = **87** and **4n-1** = 43 → **(27n+23)/2** = 592 → 296 → 148 → **74** < 87

n = 15 → 8n-1 = **119** and **4n-1** = 59 → **(27n+23)/2** = 808 → 404 → 202 → **101** < 119

... ...

n = **4n-1** → 8n-1 = **32n-9** and **4n-1** = **16n-5** → **(27(16n-5)+23)/2** = 216n-56 → 108n-28 → 54n-14 → **27n-7** < 32n-9

... ...

Further analysis on the cycles of the connections of odd numbers in D_3 are carried out in note b) and note c3).

Theorem 2n+3

2n+3 ∈ D_1 = {5; 9; 13; 17; ...; **4n+1**; ...}, **3n+5** is even if n ∈ D = {1; 3; 5; 7; ...; **2n-1**; ...}.

2n+3 ∈ D_2 = {11; 19; 27; 35; ...; **8n+3**; ...}, **(9n+16)/2** it is at least doubly even if n ∈ P_1 = {4; 8; 12; 16; ...; **4n**; ...}. This happens if (9(**4n**)+16)/2 = 18n+8 → 9n+4 is even, and this happens if n ∈ P = {2; 4; 6; 8; ...; **2n**; ...}.

2n+3 ∈ D_3 = {7; 15; 23; 31; ...; **8n-1**; ...}, **(27n+50)/2** is at least triply even if n ∈ P_2 = {2; 6; 10; 14; ...; **4n-2**;

…}. This happens if $(27(4n-2)+50)/2 = 54n-2 \rightarrow 27n-1$ is doubly even, and this happens if $n \in D_4 = \{3; 7; 11; 15; …; \mathbf{4n\text{-}1}; …\}$.

The examples given for the theorem 2n+1 also apply to the theorem 2n+3. As we will see in the section "Appendix" which formulas to apply, either derived from one or from the other theorem, is irrelevant. Here we simply validate the assertion in general cases.

$n = \mathbf{2n\text{-}1} \rightarrow 2n+3 = \mathbf{4n+1}$ and $\mathbf{3(2n\text{-}1)+5} = 6n+2 \rightarrow \mathbf{3n+1} < 4n+1$

$n = \mathbf{2n} \rightarrow 8n+3 = \mathbf{16n+3}$ and $\mathbf{4n} = \mathbf{8n} \rightarrow (9(8n)+16)/2 = 36n+8 \rightarrow 18n+4 \rightarrow \mathbf{9n+2} < 16n+3$

$n = \mathbf{4n\text{-}1} \rightarrow 8n-1 = \mathbf{32n\text{-}9}$ and $\mathbf{4n\text{-}2} = \mathbf{16n\text{-}6} \rightarrow (27(16n\text{-}6)+50)/2 = 216n-56 \rightarrow 108n-28 \rightarrow 54n-14 \rightarrow \mathbf{27n\text{-}7} < 32n-9$

b)
Demonstrations have identified three sets of odd numbers.
$D_1 = \{5; 9; 13; 17; 21; 25; 29; 33; … ; 4n+1; 4n+5; …\}$
$D_2 = \{11; 19; 27; 35; 43; 51; 59; 67; …; 8n+3; 8n+11; …\}$
$D_3 = \{7; 15; 23; 31; 39; 47; 55; 63; …; 8n-1; 8n+7; …\}$

We now verify the theorems for some values in D_1:

$5 \rightarrow 16 \rightarrow 8 \rightarrow 4 < 5$, the cycle of 5 is now linked to that of 4.
$9 \rightarrow 28 \rightarrow 14 \rightarrow 7 < 9$, the cycle of 9 is now linked to that

of 7 in D_3.

$13 \to 40 \to 20 \to 10 < 13$, the cycle of 13 is now linked to that of 10.

$17 \to 52 \to 26 \to 13 < 17$, the cycle of 17 is now linked to that of 13 in D_1.

$21 \to 64 \to 32 \to 16 < 21$, the cycle of 21 is now linked to that of 16.

$25 \to 76 \to 38 \to 19 < 25$, the cycle of 25 is now linked to that of 19 in D_2.

$29 \to 88 \to 44 \to 22 < 29$, the cycle of 29 is now linked to that of 22.

$33 \to 100 \to 50 \to 25 < 33$, the cycle of 33 is now linked to that of 25 in D_1.

$37 \to 112 \to 56 \to 28 < 37$, the cycle of 37 is now linked to that of 28.

$41 \to 124 \to 62 \to 31 < 41$, the cycle of 41 is now linked to that of 31 in D_3.

$45 \to 136 \to 68 \to 34 < 45$, the cycle of 45 is now linked to that of 34.

$49 \to 148 \to 74 \to 37 < 49$, the cycle of 49 is now linked to that of 37 in D_1.

$53 \to 160 \to 80 \to 40 < 53$, the cycle of 53 is now linked to that of 40.

$57 \to 172 \to 86 \to 43 < 57$, the cycle of 57 is now linked to that of 43 in D_2.

$61 \to 184 \to 92 \to 46 < 61$, the cycle of 61 is now linked to that of 46.

$65 \to 196 \to 98 \to 49 < 65$, the cycle of 65 is now linked to that of 49 in D_1.

$69 \rightarrow 208 \rightarrow 104 \rightarrow 52 < 69$, the cycle of 69 is now linked to that of 52.

$73 \rightarrow 220 \rightarrow 110 \rightarrow 55 < 73$, the cycle of 73 is now linked to that of 55 in D_3.

... ...

We observe that for the odd numbers in D_1 the connections of the cycles alternate between odd and even, and odd numbers alternate respecting the alternation: D_1, D_2, D_1, D_3, D_1, ...

We now verify the theorems for some values in D_2 :

$11 \rightarrow 34 \rightarrow 17 \rightarrow 52 \rightarrow 26 \rightarrow 13 \rightarrow 40 \rightarrow 20 \rightarrow 10 < 11$, the cycle of 11 is now linked to that of 10. It is noteworthy that $40 = 4 \cdot 10$.

$19 \rightarrow 58 \rightarrow 29 \rightarrow 88 \rightarrow 44 \rightarrow 22 \rightarrow 11 < 19$, the cycle of 19 is now linked to that of 11 in D_2. It is noteworthy that $44 = 4 \cdot 11$.

$27 \rightarrow 82 \rightarrow 41 \rightarrow 124 \rightarrow 62 \rightarrow 31$ (from this value is hooked to the cycle of 31 in D_3, which, as we shall see later, it connects to the cycle of 23, therefore $23 < 27$), the cycle of 27 is now linked to that of 23 in D_3.

$35 \rightarrow 106 \rightarrow 53 \rightarrow 160 \rightarrow 80 \rightarrow 40 \rightarrow 20 < 35$, the cycle of 35 is now linked to that of 20. It is noteworthy that $80 = 4 \cdot 20$.

$43 \rightarrow 130 \rightarrow 65 \rightarrow 196 \rightarrow 98 \rightarrow 49 \rightarrow 148 \rightarrow 74 \rightarrow 37 < 43$, the cycle of 43 is now linked to that of 37 in D_1. It is noteworthy that $148 = 4 \cdot 37$.

$51 \rightarrow 154 \rightarrow 77 \rightarrow 232 \rightarrow 116 \rightarrow 58 \rightarrow 29 < 51$, the cycle of 51 is now linked to that of 29 in D_1. It is noteworthy that $116 = 4 \cdot 29$.

$59 \to 178 \to 89 \to 268 \to 134 \to 67 \to 202 \to 101 \to 304 \to 152 \to 76 \to 38 < 59$, the cycle of 59 is now linked to that of 38. It is noteworthy that $152 = 4 \cdot 38$.

$67 \to 202 \to 101 \to 304 \to 152 \to 76 \to 38 < 67$, the cycle of 67 is now linked to that of 38. It is noteworthy that $152 = 4 \cdot 38$. We observe that the cycle of 67 is already present in that of 59.

$75 \to 226 \to 113 \to 340 \to 170 \to 85 \to 256 \to 128 \to 64 < 75$, the cycle of 75 is now linked to that of 64. It is noteworthy that $256 = 4 \cdot 64$.

$83 \to 250 \to 125 \to 376 \to 188 \to 94 \to 47 < 83$, the cycle of 83 is now linked to that of 47 in D_3. It is noteworthy that $188 = 4 \cdot 47$.

$91 \to 274 \to 137 \to \ldots \to 7288 \to \ldots \to 9232 \to \ldots \to 976 \to 488 \to 244 \to 122 \to 61 < 91$, the cycle of 91 is now linked to that of 61 in D_1 after 74 steps. It is noteworthy that $244 = 4 \cdot 61$.

$99 \to 298 \to 149 \to 448 \to 224 \to 112 \to 56 < 99$, the cycle of 99 is now linked to that of 56. It is noteworthy that $224 = 4 \cdot 56$.

$107 \to 322 \to 161 \to 484 \to 242 \to 121 \to 364 \to 182 \to 91 < 107$, the cycle of 107 is now linked to that of 91 in D_2. It is noteworthy that $364 = 4 \cdot 91$.

$115 \to 346 \to 173 \to 520 \to 260 \to 130 \to 65 < 115$, the cycle of 115 is now linked to that of 65 in D_1. It is noteworthy that $260 = 4 \cdot 65$.

$123 \to 370 \to 185 \to 556 \to 278 \to 139 \to 418 \to 209$

$\rightarrow 628 \rightarrow 314 \rightarrow 157 \rightarrow 472 \rightarrow 236 \rightarrow 118 < 123$, the cycle of 123 is now linked to that of 118. It is noteworthy that $472 = 4 \cdot 118$.

$131 \rightarrow 394 \rightarrow 197 \rightarrow 592 \rightarrow 296 \rightarrow 148 \rightarrow 74 < 131$, the cycle of 131 is now linked to that of 74. It is noteworthy that $296 = 4 \cdot 74$.

... ...

In the connections of the cycles of odd numbers in D_2 a precise rule is not followed. When connecting to a cycle of even numbers two odd ones may follow, to connect to a series of odd numbers two even ones may follow. The connection to a cycle of even numbers can be double, as in the case of 38 to which are connected the numbers 59 and 67. There are many other possible alternations.

We now verify the theorems for some values in D_3:

$7 \rightarrow 22 \rightarrow 11 \rightarrow 34 \rightarrow 17 \rightarrow 52 \rightarrow 26 \rightarrow 13 \rightarrow 40 \rightarrow 20 \rightarrow 10 \rightarrow 5 < 7$, the cycle of 7 is now linked to that of 5 in D_1. It is noteworthy that $40 = 8 \cdot 5$.

$15 \rightarrow 46 \rightarrow 23 \rightarrow 70 \rightarrow 35 \rightarrow 106 \rightarrow 53 \rightarrow 160 \rightarrow 80 \rightarrow 40 \rightarrow 20 \rightarrow 10 < 15$, the cycle of 15 is now linked to that of 10. It is noteworthy that $80 = 8 \cdot 10$.

$23 \rightarrow 70 \rightarrow 35 \rightarrow 106 \rightarrow 53 \rightarrow 160 \rightarrow 80 \rightarrow 40 \rightarrow 20 < 23$, the cycle of 23 is now linked to that of 20. It is noteworthy that $160 = 8 \cdot 20$.

$31 \rightarrow 94 \rightarrow 47 \rightarrow 142 \rightarrow 71 \rightarrow 214 \rightarrow 107 \rightarrow 322 \rightarrow 161 \rightarrow 484 \rightarrow 242 \rightarrow 121 \rightarrow \dots \rightarrow 182 \rightarrow 91 \rightarrow \dots \rightarrow$

$61 \rightarrow 184 \rightarrow 92 \rightarrow 46 \rightarrow 23 < 31$, the cycle of 31 is now linked to that of 23 in D_3 after 91 steps. It is noteworthy that $184 = 8 \cdot 23$.

$39 \rightarrow 118 \rightarrow 59 \rightarrow 178 \rightarrow 89 \rightarrow 268 \rightarrow 134 \rightarrow 67 \rightarrow 202 \rightarrow 101 \rightarrow 304 \rightarrow 152 \rightarrow 76 \rightarrow 38 < 39$, the cycle of 39 is now linked to that of 38. It is noteworthy that $304 = 8 \cdot 38$.

$47 \rightarrow 142 \rightarrow 71 \rightarrow 214 \rightarrow 107 \rightarrow \dots \rightarrow 91 \rightarrow \dots \rightarrow 61 \rightarrow 184 \rightarrow 92 \rightarrow 46 < 47$, the cycle of 47 is now linked to that of 46 after 88 steps. It is noteworthy that $184 = 4 \cdot 46$ as 61 is in D_1.

$55 \rightarrow 166 \rightarrow 83 \rightarrow 250 \rightarrow 125 \rightarrow 376 \rightarrow 188 \rightarrow 94 \rightarrow 47 < 55$, the cycle of 55 is now linked to that of 47 in D_3. It is noteworthy that $376 = 8 \cdot 47$.

$63 \rightarrow 190 \rightarrow 95 \rightarrow 286 \rightarrow 143 \rightarrow \dots \rightarrow 91 \rightarrow \dots \rightarrow 9232 \rightarrow \dots \rightarrow 1300 \rightarrow 650 \rightarrow 325 \rightarrow 976 \rightarrow 488 \rightarrow 244 \rightarrow 122 \rightarrow 61 < 63$, the cycle of 63 is now linked to that of 61 in D_1 after 88 steps. It is noteworthy that $488 = 8 \cdot 61$.

$71 \rightarrow 214 \rightarrow 107 \rightarrow \dots \rightarrow 91 \rightarrow \dots \rightarrow 976 \rightarrow 488 \rightarrow 244 \rightarrow 122 \rightarrow 61 < 71$, the cycle of 71 is too now linked to that of 61 in D_1 after 83 steps. It is noteworthy that $488 = 8 \cdot 61$.

$79 \rightarrow 238 \rightarrow 119 \rightarrow 358 \rightarrow 179 \rightarrow 538 \rightarrow 269 \rightarrow 808 \rightarrow 404 \rightarrow 202 \rightarrow 101 \rightarrow 304 \rightarrow 152 \rightarrow 76 < 79$, the cycle of 79 is now linked to that of 76. It is noteworthy that $304 = 4 \cdot 76$ and $808 = 8 \cdot 101$.

$87 \rightarrow 262 \rightarrow 131 \rightarrow 394 \rightarrow 197 \rightarrow 592 \rightarrow 296 \rightarrow 148 \rightarrow 74 < 87$, the cycle of 87 is now linked to that of 74. It is noteworthy that 592 = 8·74.

95 → 286 → 143 → 430 → 215 → 646 → 323 → 970 → 485 → 1456 → 728 → 364 → 182 → 91 < 95, the cycle of 95 is now linked to that of 91 in D_2. It is noteworthy that 728 = 8·91.

103 → 310 → 155 → 466 → 233 → 700 → 350 → ... → 334 → 167 → ... → 2158 → ... → 1300 → 650 → 325 → 976 → 488 → 244 → 122 → 61 < 103, the cycle of 103 is now linked to that of 61 in D_1 after 67 steps. It is noteworthy that 488 = 8·61.

111 → 334 → 167 → 502 → ... → 488 → 244 → 122 → 61 < 111, the cycle of 111 is now linked to that of 61 in D_1 after 49 steps. It is noteworthy that 488 = 8·61.

Also in the connections of the cycles of odd numbers in D_3 there is no precise rule. As we will see in the charts of links the alternations are various and unpredictable, but it is interesting to note that there are double links to the cycle of 61 for numbers 63 and 71, 103 and 111; and that this happens when the cycles are quite long.

c)
c1) Let's consider the generic odd number 4n+1 in D_1 and apply to it the algorithm of Collatz: 4n+1 → 3(4n+1)+1 = 12n+4 → 6n+2 → **3n+1** < 4n+1.

Examples:
If n = 14 → 4n+1 = 57 and 3n+1 = 43 < 57.
If n = 15 → 4n+1 = 61 and 3n+1 = 46 < 61.

... ...
If n = 20 → 4n+1 = 81 and 3n+1 = 61 < 81.
If n = 21 → 4n+1 = 85 and 3n+1 = 64 < 85.
... ...
If n = 125 → 4n+1 = 501 and 3n+1 = 376 < 501.
... ...

c2) Let's consider the generic odd number $8n+3$ in D_2 and let's apply to it the algorithm of Collatz: $8n+3 \to 3(8n+3)+1 = 24n+10 \to 12n+5$, which being odd in D_1: $12n+5 \to 3(12n+5)+1 = 36n+16 \to 18n+8 \to \mathbf{9n+4}$, which can be even or odd.
If $\mathbf{9n+4}$ is even (and it is for n even): $9n+4 \to (9n+4)/2 < 8n+3$.
If $\mathbf{9n+4}$ is odd (and it is for n odd) it connects to the cycle of an odd or in D_1 or in D_2 or in D_3.

Examples:
If n = 6 → $8n+3 = 51$ and $12n+5 = 77$ in D_1 and $9n+4 = 58 \to 29 < 51$.
If n = 7 → $8n+3 = 59$ and $12n+5 = 89$ in D_1 and $9n+4 = 67$ in $D_2 \to 202 \to 101$ in $D_1 \to 304 \to 152 \to 76 \to 38 < 59$.
If n = 8 → $8n+3 = 67$ and $12n+5 = 101$ in D_1 and $9n+4 = 76 \to 38 < 67$.
If n = 9 → $8n+3 = 75$ and $12n+5 = 113$ in D_1 and $9n+4 = 85$ in $D_1 \to 256 \to 128 \to 64 < 75$.
If n = 10 → $8n+3 = 83$ and $12n+5 = 125$ in D_1 and $9n+4 = 94 \to 47 < 83$.

If n = 11 → 8n+3 = 91 and 12n+5 = 137 in D_1 and 9n+4 = 103 in D_3 → 310 → 155 in D_2 → 466 → 233 in D_1 → ... → 319 in D_3 → ... → 325 in D_1 → 976 → 488 → 244 → 122 → 61 < 91.

... ...

If n = 26 → 8n+3 = 211 and 12n+5 = 317 in D_1 and 9n+4 = 238 → 119 < 211.

... ...

c3) Let's consider the generic odd number 8n-1 in D_3 and let's apply to it the algorithm of Collatz: 8n-1 → 3(8n-1)+1 = 24n-2 → **12n-1**, which is an odd number either in D_2 or in D_3. So we apply the algorithm again: 12n-1 → 3(12n-1)+1 = 36n-2 → **18n-1**, which is an odd either in D_1 or in D_2 or in D_3. We can therefore conclude that the cycle of an odd number in D_3 connects to the cycle of another odd number either in D_1 or in D_2 or in D_3, and then, as shown in c1) and c2), so sooner or later, there is an element in the cycle that is at least triply even that becomes less in value than its generator 8n-1.

Examples:
If n = 5 → 8n-1 = 39 and 12n-1 = 59 in D_2 and 18n-1 = 89 in D_1 → 268 → 134 → 67 in D_2 → 202 → 101 in D_1 → 304 → 152 → 76 → 38 < 39.
If n = 6 → 8n-1 = 47 and 12n-1 = 71 in D_3 and 18n-1 = 107 in D_2 → 322 → 161 in D_1 → ... → 91 in D_2 → ... → 103 in D_3 → ... → 61 in D_1 → 184 → 92 → 46 < 47.
If n = 7 → 8n-1 = 55 and 12n-1 = 83 in D_2 and 18n-1 =

125 in D_1 → 376 → 188 → 94 → 47 < 55

… …

If n = 10 → 8n-1 = 79 and 12n-1 = 119 in D_3 and 18n-1 = 179 in D_2 → 538 → 269 in D_1 → 808 → 404 → 202 → 101 in D_1 → 304 → 152 → 76 < 79.

If n = 11 → 8n-1 = 87 and 12n-1 = 131 in D_2 and 18n-1 = 197 in D_1 → 592 → 296 → 148 → 74 < 87

If n = 12 → 8n-1 = 95 and 12n-1 = 143 in D_3 and 18n-1 = 215 in D_3 → 646 → 323 in D_2 → 970 → 485 in D_1 → 1456 → 728 → 364 → 182 → 91 < 95.

… …

The formulas obtained in c1), c2) and c3), allows us to find the element from which any odd number becomes less in value than itself in the same cycle it generated. The procedure is explained in the following examples.

149 is in D_1 for n = 37 → 3n+1 = **112** < 149.

147 is in D_2 for n = 18 → 9n+4 = 166 → **83** < 147.

143 is in D_3 for n = 18 → 18n-1 = 323 in D_2 for n = 40 → 9n+4 = 364 → 182 → **91** < 143.

155 is in D_2 for n = 19 → 9n+4 = 175 in D_3 for n = 22 → 18n-1 = 395 in D_2 for n = 49 → 9n+4 = 445 in D_1 for n = 111 → 3n+1 = 334 → 167 in D_3 for n = 21 → 18n-1 = 377 in D_1 for n = 94 → 3n+1 = 283 in D_1 for n = 35 → 9n+4 = 319 in D_3 for n = 40 → 18n-1 = 719 in D_3 for n = 90 → 18n-1 = 1619 in D_2 for n = 202 → 9n+4 = 1822 → 911 in D_3 for n = 114 → 18n-1 = 2051 in D_2 for n = 256 → 9n+4 = 2308 → 1154 → 577 in D_1 for n = 144 →

$3n+1 = 433$ in D_1 for n = 108 → $3n+1 = 325$ in D_1 for n = 81 → $3n+1 = 244$ → **122** < 155.

151 is in D_3 for n = 19 → $18n-1 = 341$ in D_1 for n = 85 → $3n+1 = 256$ → **128** < 151.

163 is in D_2 for n = 20 → $9n+4 = 184$ → **92** < 163.

159 is in D_3 for n = 20 → $18n-1 = 359$ in D_3 for n = 45 → $18n-1 = 809$ in D_1 for n = 202 → $3n+1 = 607$ in D_3 for n = 76 → $18n-1 = 1637$ in D_3 for n = 171 → $18n-1 = 3077$ in D_1 for n = 769 → $3n+1 = 2308$ → 1154 → 577 in D_1 for n = 144 → $3n+1 = 433$ in D_1 for n = 108 → $3n+1 = 325$ in D_1 for n = 81 → $3n+1 = 244$ → **122** < 159.

... ...

The labyrinth of Syracuse

We have seen how the endless succession generated by the algorithm of Collatz are an inextricable labyrinth of rooms numbered in which it seems impossible to find the exit. However, with a little patience in making the calculations, the theorems demonstrated allow us to always find the exit, starting from any room in this maze that we will call: *labyrinth of Syracuse.*

In Fig. 11 shows the strategy to exit the maze of Syracuse as we stand in room number 104..

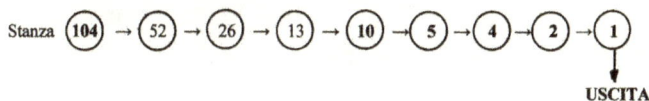

Stanza (104) → (52) → (26) → (13) → (10) → (5) → (4) → (2) → (1)
↓
USCITA

Fig. 11

In Fig. 12 shows the strategy to exit the maze of Syracuse as we stand in room number 196.

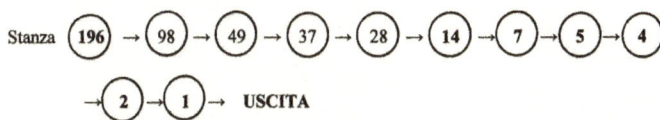

Stanza (196) → (98) → (49) → (37) → (28) → (14) → (7) → (5) → (4)
→ (2) → (1) → **USCITA**

Fig. 12

In Fig. 13 shows the strategy to exit the maze of Syracuse as we stand in room number 248.

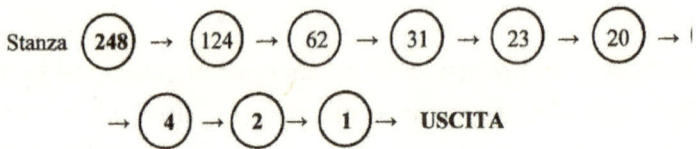

Stanza $(248) \rightarrow (124) \rightarrow (62) \rightarrow (31) \rightarrow (23) \rightarrow (20) \rightarrow$

$\rightarrow (4) \rightarrow (2) \rightarrow (1) \rightarrow$ USCITA

Fig. 13

In Fig. 14 shows the strategy to exit the maze of Syracuse as we stand in room number 1008.

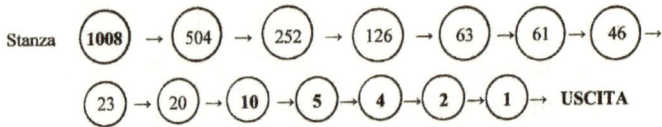

Stanza $(1008) \rightarrow (504) \rightarrow (252) \rightarrow (126) \rightarrow (63) \rightarrow (61) \rightarrow (46) \rightarrow$

$(23) \rightarrow (20) \rightarrow (10) \rightarrow (5) \rightarrow (4) \rightarrow (2) \rightarrow (1) \rightarrow$ USCITA

Fig. 14

In Fig. 15 shows the strategy to exit the maze of Syracuse as we stand in room number 99.

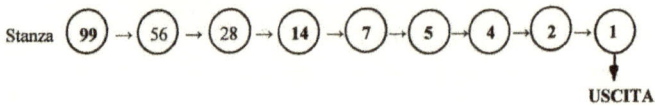

Stanza $(99) \rightarrow (56) \rightarrow (28) \rightarrow (14) \rightarrow (7) \rightarrow (5) \rightarrow (4) \rightarrow (2) \rightarrow (1)$
$\downarrow$
USCITA

Fig. 15

In Fig. 16 shows the strategy to exit the maze of Syracuse as we stand in room number 101

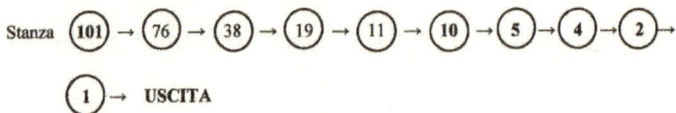

Stanza $(101) \rightarrow (76) \rightarrow (38) \rightarrow (19) \rightarrow (11) \rightarrow (10) \rightarrow (5) \rightarrow (4) \rightarrow (2) \rightarrow$

$(1) \rightarrow$ USCITA

Fig. 16

In Fig. 17 shows the strategy to exit the maze of Syracuse as we stand in room number 257.

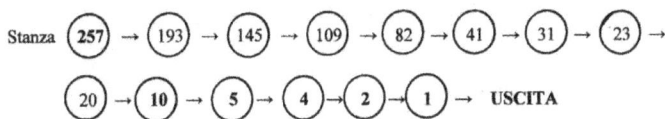

Stanza (257) → (193) → (145) → (109) → (82) → (41) → (31) → (23) →

(20) → (10) → (5) → (4) → (2) → (1) → USCITA

Fig. 17

In Fig. 18 shows the strategy to exit the maze of Syracuse as we stand in room number 1947.

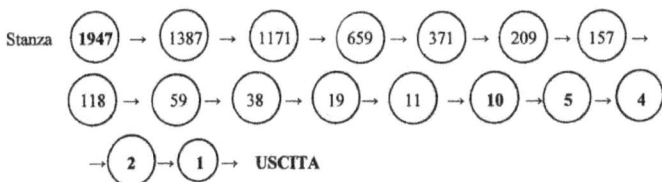

Stanza (1947) → (1387) → (1171) → (659) → (371) → (209) → (157) →

(118) → (59) → (38) → (19) → (11) → (10) → (5) → (4)

→ (2) → (1) → USCITA

Fig. 18

The solution of the conjecture of Syracuse allows us to find the exit of its endless maze beginning from any room present in it.

Annotation:
In all cycles of links, the final cycle (the last five elements = four steps) to exit the maze of Syracuse is: $\{10; 5; 4; 2; 1\}$; or, more rarely, $\{7; 5; 4; 2; 1\}$; unless in the cycle appears an even element like that of the kind of 2^p, or an odd element of the type of $(2^p-1)/3$. In this case the subsequent steps are evaluable a priori, and the last four steps are: $16 \rightarrow 8 \rightarrow 4 \rightarrow 2 \rightarrow 1 = \{16; 8; 4; 2; 1\}$.

Another final cycle in cycles of links is {12; 6; 3; 2; 1}. This cycle is anomalous, since it appears only if the generating numbers are of the type n = $3 \cdot 2^p \in I(3)$ = {3; 6; 12; 24; 48; 96; 192; 384; …; $3 \cdot 2^p$; $3 \cdot 2^{(p+1)}$; …}. The n $\in$ I(3) do not appear in any other iterative cycle. Not in those generated by even numbers, in fact: if n $\in$ P, then n $\rightarrow$ n/2, if n/2 = $3 \cdot 2^p$ is n = $2 \cdot 3 \cdot 2^p$ = $3 \cdot 2^{(p+1)} \in I(3)$. Not in those generated by odd numbers, in fact if n $\in$ D then n $\rightarrow$ 3n+1, if 3n+1 = $3 \cdot 2^p$ is n = $(3 \cdot 2^p - 1)/3$ = $2^p - 1/3$ which is impossible.

The only odd number in I(3) is 3, which, as already seen, generates the sequence: **3** $\rightarrow$ 10 $\rightarrow$ 5 $\rightarrow$ 16 $\rightarrow$ 8 $\rightarrow$ 4 $\rightarrow$ **2** < 3. In the labyrinth of Syracuse this anomalous cycle, up to the value 3, can be considered as an infinite sequence of even rooms, numbered one double of the other, aligned in a long hallway perfectly straight that is not connected to any other room of the maze. In the metaphor of the descent to the sea described in the following paragraph, however, it can be interpreted in such a river that flows into the sea without having any tributaries or being affluent. For n $\geq$ 3 all sequences generated by the numbers n = $3 \cdot 2^p \in I(3)$ are monotone decreasing such that a_n = $a_{n-1}/2$. If, for example, the generating number is n = $3 \cdot 2^{10}$ = 3072 we have S(3072) = {3072; 1536; 768; 384; 192; 96; 48; 24; 12; 6; 3}.

Descent to the sea

If instead of using the metaphor of the labyrinth, we used that of the descent to the sea, the examples above are translated into as many graphics. We simply represent only those for even number 104 and odd numbers 99 and 101. The other cycles contain a number of steps, that although finite, is too large to be represented in a graph. It should be borne in mind that the final cycle $\{10; 5; 4; 2; 1\}$ is done in 6 steps, since the $o(5)$ passes to $o(4)$ in three steps, while the final cycle $\{7; 5; 4; 2; 1\}$ is accomplished in 16 steps, since the $o(7)$ passes to $o(5)$ in 11 steps and, as we already said, the $o(5)$ passes to $o(4)$ in three steps. For convenience these end cycles are represented by a single line section, (one step).

Figure 19 shows the graph of the descent to the sea from the main horizon of the number 104. This leads to the sea in 5 steps instead of 12.

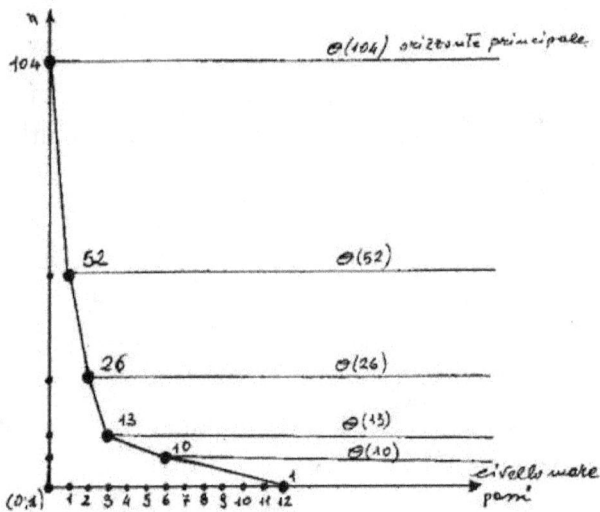

Fig. 19

Figure 20 shows the graph of the descent to the sea from the main horizon of the number 99. This leads to the sea in 5 steps instead of 25.

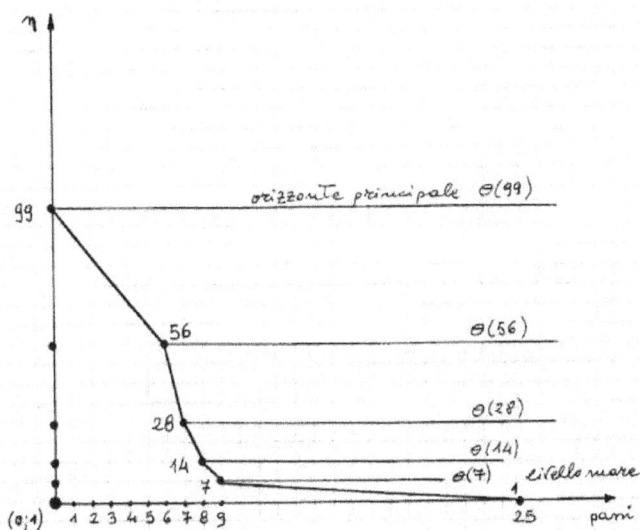

Fig. 20

Figure 21 shows the graph of the descent to the sea from the main horizon of the number 101. This leads to the sea in 6 steps instead of 25.

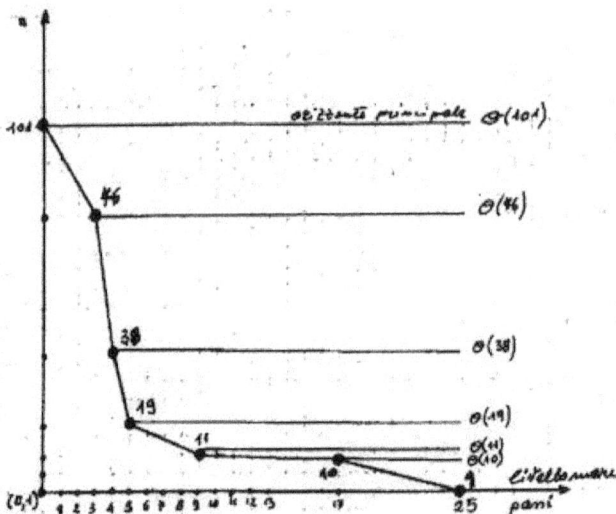

Fig. 21

The solution of the conjecture of Syracuse allows to reach the sea by the highest hight greatly reducing the number of steps and remaining always below the main horizon.

The skyscraper

If we wanted to get off on the first floor of a very high skyscraper finding ourselves on the 63rd floor, we should follow a clear strategy in pressing the buttons of the elevator, otherwise we will continue to go down and back up to even the 9232rd floor. Instead, using the method of the cycles of links, we know that we have to press the button 61 of the elevator control panel. And then the button 46. And then the button 23. Since we already know the cycle of the links of number 23, we will get off on the first floor (1) with ease and from there, with the stairs, we will reach the ground floor. If we do not follow this strategy we would be forced to stay locked in the elevator, going up and down, for over 107 floors. This strategy is summarized in Fig. 22.

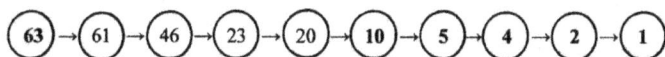

$$63 \rightarrow 61 \rightarrow 46 \rightarrow 23 \rightarrow 20 \rightarrow 10 \rightarrow 5 \rightarrow 4 \rightarrow 2 \rightarrow 1$$

Fig. 22

Conclusion

Having proved the conjecture of Syracuse, we can state the following theorem.

Theorem of Collatz

If at any natural integer n, non-zero, we will apply the algorithm 3n+1 if n is odd, n/2 if n is even, the sequence of the values thus obtained precipitates to 1 after a finite number of steps, always in compliance with the final cycle {4; 2; 1}.

Appendix

The proofs of the theorems that solve the conjecture of Syracuse offer many possible applications. Let's analyses some of them.

1)
In any cycle it's possible to compute the element from which it becomes less in value of its generator.

We apply the formulas of the theorem 2n+3.

201 = 4n+1 for n = 50, then 201 is in D_1. To n = 50 corresponds the odd number 2n-1 = 99 in D. Hence 3n+5 = 302 → **156** < 201.

83 = 8n+3 for n = 10, then 83 is in D_2. To n = 10 corresponds the even number 4n = 40 in P_1. Hence (9n+16)/2 = 188 → 94 → **47** < 83.

119 = 8n-1 for n = 15, then 119 is in D_3. To n = 15 corresponds the even number 4n-2 = 58 in P_2. Hence (27n+50)/2 = 808 → 404 → 202 → **101** < 119.

We apply the formulas of the theorem 2n+1.

201 = 4n+1 for n = 50, then 201 is in D_1. To n = 50 corresponds the even number 2n = 100 in P. Hence 3n+2 = 302 → **156** < 201.

83 = 8n+3 for n = 10, then 83 is in D_2. To n = 10 corresponds the odd number 4n+1 = 41 in D_1. Hence (9n+7)/2

= 188 → 94 → **47** < 83.

119 = 8n-1 for n = 15, then 119 is in D_3. To n = 15 corresponds the odd number 4n-1 = 59 in D_4. Hence (27n+23)/2 = 808 → 404 → 202 → **101** < 119.

The results achieved are the same.

2)

∀ n ∈ N can be traced back to the corresponding odd number in D_1, D_2, D_3, and calculate the generating number of the cycle to which it connects.

We apply the formulas of the theorem 2n+3.

For n = 10, 4n+1 = 41 in D_1 and 2n-1 = 19 in D, then 3n+5 = 62 → 31 < 41.

For n = 10, 8n+3 = 83 in D_2 and 4n = 40 in P_1, then (9n+16)/2 = 188 → 94 → 47 < 83.

For n = 10, 8n-1 = 79 in D_3 and 4n-2 = 38 in P_2, then (27n+50)/2 = 269 in D_1 for n = 67 and 2n-1 = 133 in D, then 3n+5 = 404 → 202 → 101 in D_1 for n = 25 and 2n-1 = 49 in D, then 3n+5 = 152 → 76 < 79.

For n = 11, 4n+1 = 45 in D_1 and 2n-1 = 21 in D, then 3n+5 = 68 → 34 < 45.

For n = 11, 8n + 3 = 91 in D_2 and 4n = 44 in P_1, then (9n+16)/2 = 206 → 103 in D_3 for n = 13 and 4n-2 = 50 in P_2, then $(27n+50)/2 = 700 \rightarrow 350 \rightarrow 175$ in D_3 for n = 22 and 4n-2 = 86 in P_2, then $(27n+50)/2 = 1186 \rightarrow 593$ in D_1 for n = 148 and 2n-1 = 295 in D, then 3n+5 = 890 $\rightarrow 445$ in D_1 for n = 111 and 2n-1 = 221 in D, then 3n+5 = 668 $\rightarrow 334 \rightarrow 167$ in D_3 for n = 21 and 4n-2 = 82 in P_2, then $(27n+50)/2 = 1132 \rightarrow 566 \rightarrow 283$ in D_2 for n = 35 and 4n = 140 in P_1, then $(9n+16)/2 = 638 \rightarrow 319$ in D_3 for n = 40 and 4n-2 = 158 in P_2, then $(27n+50)/2 = 2158 \rightarrow 1079$ in D_3 for n = 135 and 4n-2 = 538 in P_2, then $(27n+50)/2 = 7288 \rightarrow 3644 \rightarrow 1822 \rightarrow 911$ in D_3 for n = 114 and 4n-2 = 454 in P_2, then $(27n+50)/2 = 6154$ $\rightarrow 3077$ in D_1 for n = 769 and 2n-1 = 1537 in D, then 3n+5 = 4616 $\rightarrow 2308 \rightarrow 1154 \rightarrow 577$ in D_1 for n = 144 and 2n-1 = 287 in D, then 3n+5 = 866 $\rightarrow 433$ in D_1 for n = 108 and 2n-1 = 215 in D, then 3n+5 = 650 $\rightarrow 325$ in D_1 for n = 81 and 2n-1 = 161 in D, then 3n+5 = 488 $\rightarrow$ 244 $\rightarrow 122 \rightarrow 61 < 91$.

For n = 11, 8n-1 = 87 in D_3 and 4n-2 = 42 in P_2, then: $(27n+50)/2 = 592 \rightarrow 296 \rightarrow 148 \rightarrow 74 < 87$.

The same results are obtained by applying the formulas of the theorem 2n+1.

Such a method is applicable to any odd number in D_1, D_2, D_3. For the odd numbers in D_1 the method is very simple, while for odd numbers in D_2 and in D_3 it requires some patience. In the cycles of these odd numbers, especially in very long cycles, the connections, dealt with in the explanatory notes c2) and c3), are repeated several times before the element of a cycle becomes less than its generator; as it has happened in the cycle of 91 in the previous example.

3)
With reference to paragraph "The labyrinth of Syracuse", if we wanted to get out of the maze standing in room number 101, odd, the proof of the theorem Collatz suggests the following strategy:

101 = 4n+1 in D_1 for n = 25 $\rightarrow$ 2n-1 = 49 in D $\rightarrow$ 3n+5 = 152 $\rightarrow$ **76** < 101

76 $\rightarrow$ **38** < 76

38 $\rightarrow$ **19** < 38

19 = 8n+3 in D_2 for n = 2 $\rightarrow$ 4n = 8 in P_1 $\rightarrow$ (9n+16)/2 = 44 $\rightarrow$ 22 $\rightarrow$ **11** < 19

11 = 8n+3 in D_2 for n = 1 $\rightarrow$ 4n = 4 in P_1 $\rightarrow$ (9n+16)/2 = 26 $\rightarrow$ 13 = 4n+1 in D_1 for n = 3 $\rightarrow$ 2n-1 = 5 in D $\rightarrow$ 3n+5 = 20 $\rightarrow$ **10** < 11

10 $\rightarrow$ **5** < 10

5 = 4n+1 in D_1 for n = 1 $\rightarrow$ 2n-1 = 1 in D $\rightarrow$ 3n+5 = 8 $\rightarrow$ **4** < 5

4 $\rightarrow$ **2** $\rightarrow$ **1** $\rightarrow$ **EXIT**

Finding the cycle of the connections of Fig. 16 of that paragraph:

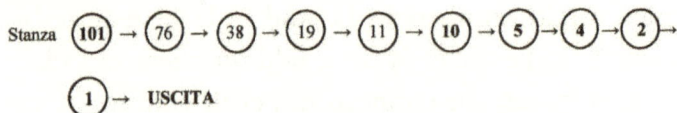

Stanza (101) $\rightarrow$ (76) $\rightarrow$ (38) $\rightarrow$ (19) $\rightarrow$ (11) $\rightarrow$ (10) $\rightarrow$ (5) $\rightarrow$ (4) $\rightarrow$ (2) $\rightarrow$

(1) $\rightarrow$ USCITA

To the cycle of links of the odd number 101, we apply the formulas derived in the demonstration of the theorem 2n+1.

101 = 4n+1 in D_1 for n = 25 → 2n = 50 in P → 3n+2 = 152 → **76** < 101
76 → **38** < 76
38 → **19** < 38
19 = 8n+3 in D_2 for n = 2 → 4n +1 = 9 in D_1 → (9n+7)/2 = 44 → 22 → **11** < 19
11 = 8n+3 in D_2 for n = 1 → 4n+1 = 5 in D_1 → (9n+7)/2 = 26 → 13 = 4n+1 in D_1 for n = 3 → 2n = 6 in P → 3n+2 = 20 → **10** < 11
10 → **5** < 10
5 = 4n+1 in D_1 for n = 1 → 2n = 2 in P → 3n+2 = 8 → **4** < 5
4 → **2** → **1** → **EXIT**

The results of the connections are perfectly identical to those obtained by applying the formulas derived in the proof of the theorem 2n+3.

We apply the formulas of the theorem 2n+1 to exit the maze of Syracuse finding ourselves in odd number room 123.

123 in D_2 for n = 15 → 4n+1 = 61 in D_1 → (9n+7)/2 = 278 → 139 in D_2 for n = 17 → 4n+1 = 69 in D_1 → (9n+7)/2 = 314 → 157 in D_1 for n = 39 → 2n = 78 in P

$\rightarrow$ 3n+2 = 236 $\rightarrow$ **118** < 123

118 $\rightarrow$ **59** < 118

59 in D_2 for n = 7 $\rightarrow$ 4n+1 = 29 in D_1 $\rightarrow$ (9n+7)/2 = 134 $\rightarrow$ 67 in D_2 for n = 8 $\rightarrow$ 4n+1 = 33 in D_1 $\rightarrow$ (9n+7)/2 = 152 $\rightarrow$ 76 $\rightarrow$ **38** < 59

38 $\rightarrow$ **19** < 38

19 in D_2 for n = 2 $\rightarrow$ 4n+1 = 9 in D_1 $\rightarrow$ (9n+7)/2 = 44 $\rightarrow$ 22 $\rightarrow$ **11** < 19

11 in D_2 for n = 1 and 4n+1 = 5 in D_1 $\rightarrow$ (9n+7)/2 = 26 $\rightarrow$ 13 in D_1 for n = 3 $\rightarrow$ 2n = 6 in P $\rightarrow$ 3n+2 = 20 $\rightarrow$ **10** < 11

10 $\rightarrow$ **5** $\rightarrow$ **4** $\rightarrow$ **2** $\rightarrow$ **1 FINAL CYCLE**

The cycle of the connections is shown in Fig. 23.

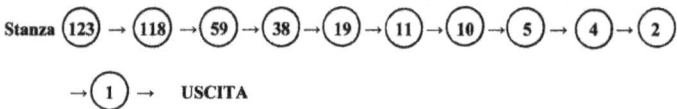

Stanza (123) → (118) → (59) → (38) → (19) → (11) → (10) → (5) → (4) → (2) → (1) → USCITA

Fig. 23

Other examples of application formulas obtained from theorem 2n+1:

1666 $\rightarrow$ 833 in D_1 for n = 208 $\rightarrow$ 2n = 416 in P $\rightarrow$ 3n+2 = 1250 $\rightarrow$ 625 in D_1 for n = 156 $\rightarrow$ 2n = 312 in P $\rightarrow$ 3n+2 = 932 $\rightarrow$ 469 in D_1 for n = 117 $\rightarrow$ 2n = 234 in P $\rightarrow$ 3n+2 = 704 $\rightarrow$ 352 $\rightarrow$ 176 $\rightarrow$ 88 $\rightarrow$ 44 $\rightarrow$ 22 $\rightarrow$ 11 $\rightarrow$ 10 $\rightarrow$ ... $\rightarrow$ Exit.

Synthesis:

1666 $\rightarrow$ 833 $\rightarrow$ 625 $\rightarrow$ 469 $\rightarrow$ 352 $\rightarrow$ 176 $\rightarrow$ 88 $\rightarrow$ 44

$\rightarrow 22 \rightarrow 11 \rightarrow 10 \rightarrow 5 \rightarrow 4 \rightarrow 2 \rightarrow 1$

2224 $\rightarrow 1112 \rightarrow 556 \rightarrow 278 \rightarrow 139$ in D_2 for n = 17 $\rightarrow$
4n+1 = 69 in $D_1 \rightarrow (9n+7)/2 = 314 \rightarrow 157$ in D_1 for n =
39 $\rightarrow$ 2n = 78 in P $\rightarrow$ 3n+2 = 236 $\rightarrow 118 \rightarrow 59$ in D_2 for
n = 7 $\rightarrow$ 4n+1 = 29 in $D_1 \rightarrow (9n+7)/2 = 134 \rightarrow 67$ in D_2
for n = 8 $\rightarrow$ 4n+1 = 33 in $D_1 \rightarrow (9n+7)/2 = 152 \rightarrow 76 \rightarrow$
38 $\rightarrow$ 19 in D_2 for n = 2 $\rightarrow$ 4n+1 = 9 in $D_1 \rightarrow (9n+7)/2$
= 44 $\rightarrow 22 \rightarrow 11$ in D_2 for n = 1 $\rightarrow$ 4n+1 = 5 in $D_1 \rightarrow$
(9n+7)/2 = 26 $\rightarrow 13$ in D_1 for n = 3 $\rightarrow$ 2n = 6 in P $\rightarrow$
3n+2 = 20 $\rightarrow 10 \rightarrow ... \rightarrow$ Exit.
Synthesis:
2224 $\rightarrow 1112 \rightarrow 556 \rightarrow 278 \rightarrow 139 \rightarrow 118 \rightarrow 59 \rightarrow 38$
$\rightarrow 19 \rightarrow 11 \rightarrow 10 \rightarrow 5 \rightarrow 4 \rightarrow 2 \rightarrow 1$

2428 $\rightarrow 1214 \rightarrow 607$ in D_3 for n = 76 $\rightarrow$ 4n-1 = 303 in
$D_4 \rightarrow (27n+23)/2 = 4102 \rightarrow 2051$ in D_2 for n = 256 $\rightarrow$
4n+1 = 1025 in $D_1 \rightarrow (9n+7)/2 = 4616 \rightarrow 2308 \rightarrow 1154$
$\rightarrow 577$ in D_1 for n = 144 $\rightarrow$ 2n = 288 in P $\rightarrow$ 3n+2 =
866 $\rightarrow 433$ in D_1 for n = 108 $\rightarrow$ 2n = 216 in P $\rightarrow$ 3n+2
= 650 $\rightarrow 325$ in D_1 for n = 81 $\rightarrow$ 2n = 162 in P $\rightarrow$ 3n+2
= 488 $\rightarrow 244 \rightarrow 122 \rightarrow 61$ in D_1 for n = 15 $\rightarrow$ 2n = 30
in P $\rightarrow$3n+2 = 92 $\rightarrow 46 \rightarrow 23$ in D_3 for n = 3 $\rightarrow$ 4n-1 =
11 in $D_4 \rightarrow (27n+23)/2 = 160 \rightarrow 80 \rightarrow 40 \rightarrow 20 \rightarrow 10$
$\rightarrow ... \rightarrow$ Exit.
Synthesis:
2428 $\rightarrow 1214 \rightarrow 607 \rightarrow 577 \rightarrow 433 \rightarrow 325 \rightarrow 244 \rightarrow$
122 $\rightarrow 61 \rightarrow 46 \rightarrow 23 \rightarrow 20 \rightarrow 10 \rightarrow 5 \rightarrow 4 \rightarrow 2 \rightarrow 1$

2888 $\rightarrow$ 1444 $\rightarrow$ 722 $\rightarrow$ 361 in D_1 for n = 90 $\rightarrow$ 2n = 180 in P $\rightarrow$ 3n+2 = 542 $\rightarrow$ 271 in D_3 for n = 34 $\rightarrow$ 4n-1 = 135 in D_4 $\rightarrow$ (27n+23)/2 = 1834 $\rightarrow$ 917 in D_1 for n = 229 $\rightarrow$ 2n = 458 in P $\rightarrow$ 3n+2 = 1376 $\rightarrow$ 688 $\rightarrow$ 344 $\rightarrow$ 172 $\rightarrow$ 86 $\rightarrow$ 43 in D_2 for n = 5 $\rightarrow$ 4n+1 = 21 in D_1 $\rightarrow$ (9n+7)/2 = 98 $\rightarrow$ 49 in D_1 for n = 12 $\rightarrow$ 2n = 24 in P $\rightarrow$ 3n+2 = 74 $\rightarrow$ 37 in D_1 for n = 9 $\rightarrow$ 2n = 18 in P $\rightarrow$ 3n+2 = 56 $\rightarrow$ 28 $\rightarrow$ 14 $\rightarrow$ ... $\rightarrow$ Exit.

Synthesis:
2888 $\rightarrow$ 1444 $\rightarrow$ 722 $\rightarrow$ 361 $\rightarrow$ 271 $\rightarrow$ 172 $\rightarrow$ 86 $\rightarrow$ 43 $\rightarrow$ 37 $\rightarrow$ 28 $\rightarrow$ 14 $\rightarrow$ 7 $\rightarrow$ 5 $\rightarrow$ 4 $\rightarrow$ 2 $\rightarrow$ 1

3220 $\rightarrow$ 1610 $\rightarrow$ 805 in D_1 for n = 201 $\rightarrow$ 2n = 402 in P $\rightarrow$ 3n+2 = 1208 $\rightarrow$ 604 $\rightarrow$ 302 $\rightarrow$ 151 in D_3 for n = 19 $\rightarrow$ 4n-1 = 75 in D_4 $\rightarrow$ (27n+23)/2 = 1024 $\rightarrow$ 512 $\rightarrow$ 256 $\rightarrow$ 128 $\rightarrow$ 64 $\rightarrow$ 32 $\rightarrow$ 16 $\rightarrow$ ... $\rightarrow$ Exit.

Synthesis:
3220 $\rightarrow$ 1610 $\rightarrow$ 805 $\rightarrow$ 604 $\rightarrow$ 302 $\rightarrow$ 151 $\rightarrow$ 128 $\rightarrow$ 64 $\rightarrow$ 32 $\rightarrow$ 16 $\rightarrow$ 8 $\rightarrow$ 4 $\rightarrow$ 2 $\rightarrow$ 1

5220 $\rightarrow$ 2610 $\rightarrow$ 1305 in D_1 for n = 326 $\rightarrow$ 2n = 652 in P $\rightarrow$ 3n+2 = 1958 $\rightarrow$ 979 in D_2 for n = 122 $\rightarrow$ 4n+1 = 489 in D_1 $\rightarrow$ (9n+7)/2 = 2204 $\rightarrow$ 1102 $\rightarrow$ 551 in D_3 for n = 69 $\rightarrow$ 4n-1 = 275 in D_4 $\rightarrow$ (27n+23)/2 = 3724 $\rightarrow$ 1862 $\rightarrow$ 931 in D_2 for n = 116 $\rightarrow$ 4n+1 = 465 in D_1 $\rightarrow$ (9n+7)/2 = 2096 $\rightarrow$ 1048 $\rightarrow$ 524 $\rightarrow$ 262 $\rightarrow$ 131 in D_2 for n = 16 $\rightarrow$ 4n+1 = 65 in D_1 $\rightarrow$ (9n+7)/2 = 296 $\rightarrow$ 148 $\rightarrow$ 74 $\rightarrow$ 37 in D_1 for n = 9 $\rightarrow$ 2n = 18 in P $\rightarrow$ 3n+2 = 56 $\rightarrow$ 28 $\rightarrow$ 14 $\rightarrow$ 7 $\rightarrow$ ... $\rightarrow$ Exit.

Synthesis:

$5220 \rightarrow 2610 \rightarrow 1305 \rightarrow 979 \rightarrow 551 \rightarrow 524 \rightarrow 262 \rightarrow$
$131 \rightarrow 74 \rightarrow 37 \rightarrow 28 \rightarrow 14 \rightarrow 7 \rightarrow 5 \rightarrow 4 \rightarrow 2 \rightarrow 1$

$\mathbf{6604} \rightarrow 3302 \rightarrow 1651$ in D_2 for n = 206 $\rightarrow 4n+1 = 825$
in $D_1 \rightarrow (9n+7)/2 = 3716 \rightarrow 1858 \rightarrow 929$ in D_1 for n =
$232 \rightarrow 2n = 464$ in P $\rightarrow 3n+2 = 1394 \rightarrow 697$ in D_1 for
n = 174 $\rightarrow 2n = 348$ in P $\rightarrow 3n+2 = 1046 \rightarrow 523$ in D_2
for n = 65 $\rightarrow 4n+1 = 261$ in $D_1 \rightarrow (9n+7)/2 = 1178 \rightarrow$
589 in D_1 for n = 147 $\rightarrow 2n = 294$ in P $\rightarrow 3n+2 = 884 \rightarrow$
$442 \rightarrow 221$ in D_1 for n = 55 $\rightarrow 2n = 110$ in P $\rightarrow 3n+2 =$
$332 \rightarrow 166 \rightarrow 83$ in D_2 for n = 10 $\rightarrow 4n+1 = 41$ in $D_1 \rightarrow$
$(9n+7)/2 = 188 \rightarrow 94 \rightarrow 47 \rightarrow \ldots \rightarrow 46 \rightarrow 23$ in D_3 for
n = 3 $\rightarrow 4n-1 = 11$ in $D_4 \rightarrow (27n+23)/2 = 160 \rightarrow 80 \rightarrow$
$40 \rightarrow 20 \rightarrow 10 \rightarrow \ldots \rightarrow$ Exit.
Synthesis:
$6604 \rightarrow 3302 \rightarrow 1651 \rightarrow 929 \rightarrow 697 \rightarrow 523 \rightarrow 442 \rightarrow$
$221 \rightarrow 166 \rightarrow 83 \rightarrow 47 \rightarrow 46 \rightarrow 23 \rightarrow 20 \rightarrow 10 \rightarrow 5$
$\rightarrow 4 \rightarrow 2 \rightarrow 1$

$\mathbf{118000} \rightarrow 59000 \rightarrow 29500 \rightarrow 14750 \rightarrow 7375$ in D_3 for
n = 922 $\rightarrow 4n-1 = 3687$ in $D_4 \rightarrow (27n+23)/2 = 49786 \rightarrow$
24893 in D_1 for n = 6223 $\rightarrow 2n = 12446$ in P $\rightarrow 3n+2 =$
$37340 \rightarrow 18670 \rightarrow 9335$ in D_3 for n = 1167 $\rightarrow 4n-1 =$
4667 in $D_4 \rightarrow (27n+23)/2 = 63016 \rightarrow 31508 \rightarrow 15754$
$\rightarrow 7877$ in D_1 for n = 1969 $\rightarrow 2n = 3938$ in P $\rightarrow 3n+2$
$= 11816 \rightarrow 5908 \rightarrow 2954 \rightarrow 1477$ in D_1 for n = 369 $\rightarrow$
2n = 738 in P $\rightarrow 3n+2 = 2216 \rightarrow 1108 \rightarrow 554 \rightarrow 277$ in
D_1 for n = 69 $\rightarrow 2n = 138$ in P $\rightarrow 3n+2 = 416 \rightarrow 208 \rightarrow$

$104 \rightarrow 52 \rightarrow 26 \rightarrow 13$ in D_1 for $n = 3 \rightarrow 2n = 6$ in $P \rightarrow$
$3n+2 = 20 \rightarrow 10 \rightarrow \dots \rightarrow$ Exit.

Synthesis:

$118000 \rightarrow 59000 \rightarrow 29500 \rightarrow 14750 \rightarrow 7375 \rightarrow 5908$
$\rightarrow 2954 \rightarrow 1477 \rightarrow 1108 \rightarrow 554 \rightarrow 277 \rightarrow 208 \rightarrow 104$
$\rightarrow 52 \rightarrow 26 \rightarrow 13 \rightarrow 10 \rightarrow 5 \rightarrow 4 \rightarrow 2 \rightarrow 1$

...

...

Conclusion

The theorems that solve the conjecture of Syracuse allow us to replace the cycles of the algorithm of Collatz with the cycles of connections, transforming their oscillating sequences in monotone decreasing sequences, which, after a finite number of steps (very low), fall to 1 always respecting the final cycles {10; 5; 4; 2; 1} or {7; 5; 4; 2; 1}.

We end with the cycles of connections of the odd number 6777 (Fig. 24) and the odd number 10131 (Fig. 25), which synthesize effectively the work done.

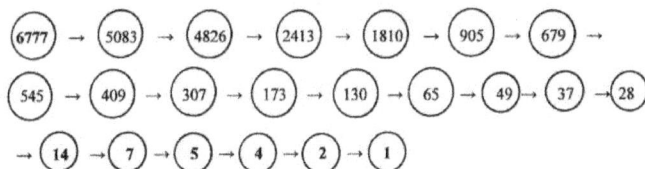

Fig. 24

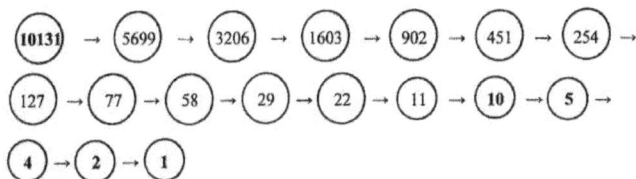

Fig. 25

Rolando Zucchini

The conjecture of Syracuse

Charts of lynks
5 - 1999

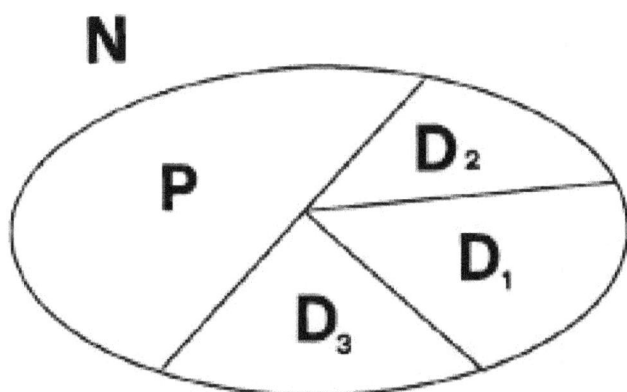

(5 – 99)

D_1	D_2	D_3
$5 \rightarrow 4$	$11 \rightarrow 10$	$7 \rightarrow 5\ D_1$
$9 \rightarrow 7\quad D_3$	$19 \rightarrow 11\ D_2$	$15 \rightarrow 10$
$13 \rightarrow 10$	$27 \rightarrow 23\ D_3$	$23 \rightarrow 20$
$17 \rightarrow 13\ D_1$	$35 \rightarrow 20$	$31 \rightarrow 23\ D_3$
$21 \rightarrow 16$	$43 \rightarrow 37\ D_1$	$39 \rightarrow 38$
$25 \rightarrow 19\ D_2$	$51 \rightarrow 29\ D_1$	$47 \rightarrow 46$
$29 \rightarrow 22$	$59 \rightarrow 38$	$55 \rightarrow 47\ D_3$
$33 \rightarrow 25\ D_1$	$67 \rightarrow 38$	$63 \rightarrow 61\ D_1$
$37 \rightarrow 28$	$75 \rightarrow 64$	$71 \rightarrow 61\ D_1$
$41 \rightarrow 31\ D_3$	$83 \rightarrow 47\ D_3$	$79 \rightarrow 76$
$45 \rightarrow 34$	$91 \rightarrow 61\ D_1$	$87 \rightarrow 74$
$49 \rightarrow 37\ D_1$	$99 \rightarrow 56$	$95 \rightarrow 91\ D_2$
$53 \rightarrow 40$		
$57 \rightarrow 43\ D_2$		
$61 \rightarrow 46$		
$65 \rightarrow 49\ D_1$		
$69 \rightarrow 52$		
$73 \rightarrow 55\ D_3$		
$77 \rightarrow 58$		
$81 \rightarrow 61\ D_1$		
$85 \rightarrow 64$		
$89 \rightarrow 67\ D_2$		
$93 \rightarrow 70$		
$97 \rightarrow 73\ D_1$		

(101 – 199)

D_1

101 → 76
105 → 79 D_3
109 → 82
113 → 85 D_1
117 → 88
121 → 91 D_2
125 → 94
129 → 97 D_1
133 → 100
137 → 103 D_3
141 → 106
145 → 109 D_1
149 → 112
153 → 115 D_2
157 → 118
161 → 121 D_1
165 → 124
169 → 127 D_3
173 → 130
177 → 133 D_1
181 → 136
185 → 139 D_2
189 → 142
193 → 145 D_1
197 → 148

D_2

107 → 91 D_2
115 → 65 D_1
123 → 118
131 → 74
139 → 118
147 → 83 D_2
155 → 122
163 → 92
171 → 145 D_1
179 → 101 D_1
187 → 119 D_3
195 → 110

D_3

103 → 61 D_1
111 → 61 D_1
119 → 101 D_1
127 → 77 D_1
135 → 86
143 → 91 D_2
151 → 128
159 → 122
167 → 122
175 → 167 D_3
183 → 155 D_2
191 → 154
199 → 190

(**201 – 299**)

D_1	D_2	D_3
201 → 151 D_3	203 → 172	207 → 167 D_3
205 → 154	211 → 119 D_3	215 → 182
209 → 157 D_1	219 → 209 D_1	223 → 122
213 → 160	227 → 128	231 → 124
217 → 163 D_2	235 → 199 D_3	239 → 122
221 → 166	243 → 137 D_1	247 → 209 D_1
225 → 169 D_1	251 → 244	255 → 205 D_1
229 → 172	259 → 146	263 → 167 D_3
233 → 175 D_3	267 → 226	271 → 172
237 → 178	275 → 155 D_2	279 → 236
241 → 181 D_1	283 → 244	287 → 205 D_1
245 → 184	291 → 164	295 → 281 D_1
249 → 187 D_2	299 → 253 D_1	
253 → 190		
257 → 193 D_1		
261 → 196		
265 → 199 D_3		
269 → 202		
273 → 205 D_1		
277 → 208		
281 → 211 D_2		
285 → 214		
289 → 217 D_1		
293 → 220		
297 → 223 D_3		

.

(301 – 399)

D_1	D_2	D_3
301 → 226	307 → 173 D_1	303 → 244
305 → 229 D_1	315 → 200	311 → 263 D_3
309 → 232	323 → 182	319 → 244
313 → 235 D_2	331 → 280	327 → 250
317 → 238	339 → 191 D_3	335 → 319 D_3
321 → 241 D_1	347 → 248	343 → 290
325 → 244	355 → 200	351 → 334
329 → 247 D_3	363 → 307 D_2	359 → 325 D_1
333 → 250	371 → 209 D_1	367 → 262
337 → 253 D_1	379 → 361 D_1	375 → 317 D_1
341 → 256	387 → 218	383 → 205 D_1
345 → 259 D_2	395 → 334	391 → 248
349 → 262		399 → 253 D_1
353 → 265 D_1		
357 → 268		
361 → 271 D_3		
365 → 274		
369 → 277 D_1		
373 → 280		
377 → 283 D_2		
381 → 286		
385 → 289 D_1		
389 → 292		
393 → 295 D_3		
397 → 298		

.

(401 – 499)

D_1

401 → 301 D_1
405 → 304
409 → 307 D_2
413 → 310
417 → 313 D_1
421 → 316
425 → 319 D_3
429 → 322
433 → 325 D_1
437 → 328
441 → 331 D_2
445 → 334
449 → 337 D_1
453 → 340
457 → 343 D_3
461 → 346
465 → 349 D_1
469 → 352
473 → 355 D_2
477 → 358
481 → 361 D_1
485 → 364
489 → 367 D_3
493 → 370
497 → 373 D_1

.

D_2

403 → 227 D_2
411 → 248
419 → 236
427 → 361 D_1
435 → 245 D_1
443 → 281 D_1
451 → 254
459 → 388
467 → 263 D_3
475 → 452
483 → 272
491 → 415 D_3
499 → 281 D_1

D_3

407 → 344
415 → 250
423 → 302
431 → 410
439 → 371 D_2
447 → 346
455 → 433 D_1
463 → 248
471 → 398
479 → 433 D_1
487 → 248
495 → 478

(**501 – 599**)

D_1	D_2	D_3
501 → 376	507 → 362	503 → 425D_1
505 → 379 D_2	515 → 290	511 → 346
509 → 382	523 → 442	519 → 329 D_1
513 → 385 D_1	531 → 299 D_2	527 → 334
517 → 388	539 → 433 D_1	535 → 452
521 → 391 D_3	547 → 308	543 → 436
525 → 394	555 → 469 D_1	551 → 524
529 → 397 D_1	563 → 317 D_1	559 → 505 D_1
533 → 400	571 → 362	567 → 479 D_3
537 → 403 D_2	579 → 326	575 → 410
541 → 406	587 → 496	583 → 416
545 → 409 D_1	595 → 335 D_3	591 → 562
549 → 412		599 → 506
553 → 415 D_3		
557 → 418		
561 → 421 D_1		
565 → 424		
569 → 427 D_2		
573 → 430		
577 → 433 D_1		
581 → 436		
585 → 439 D_3		
589 → 442		
593 → 445 D_1		
597 → 448		

.

(601 – 699)

D_1	D_2	D_3
$601 \rightarrow 451\ D_2$	$603 \rightarrow 545\ D_1$	$607 \rightarrow 577\ D_1$
$605 \rightarrow 454$	$611 \rightarrow 344$	$615 \rightarrow 329\ D_1$
$609 \rightarrow 457\ D_1$	$619 \rightarrow 523\ D_2$	$623 \rightarrow 500$
$613 \rightarrow 460$	$627 \rightarrow 353\ D_1$	$631 \rightarrow 533\ D_1$
$617 \rightarrow 463\ D_3$	$635 \rightarrow 604$	$639 \rightarrow 365\ D_1$
$621 \rightarrow 466$	$643 \rightarrow 362$	$647 \rightarrow 410$
$625 \rightarrow 469\ D_1$	$651 \rightarrow 550$	$655 \rightarrow 415\ D_3$
$629 \rightarrow 472$	$659 \rightarrow 371\ D_2$	$663 \rightarrow 560$
$633 \rightarrow 475\ D_2$	$667 \rightarrow 572$	$671 \rightarrow 346$
$637 \rightarrow 478$	$675 \rightarrow 380$	$679 \rightarrow 545\ D_1$
$641 \rightarrow 481\ D_1$	$683 \rightarrow 577\ D_1$	$687 \rightarrow 653\ D_1$
$645 \rightarrow 484$	$691 \rightarrow 389\ D_1$	$695 \rightarrow 587\ D_3$
$649 \rightarrow 487\ D_3$	$699 \rightarrow 391\ D_3$	
$653 \rightarrow 490$		
$657 \rightarrow 493\ D_1$		
$661 \rightarrow 496$		
$665 \rightarrow 499\ D_2$		
$669 \rightarrow 502$		
$673 \rightarrow 505\ D_1$		
$677 \rightarrow 508$		
$681 \rightarrow 511\ D_3$		
$685 \rightarrow 514$		
$689 \rightarrow 517\ D_1$		
$693 \rightarrow 520$		
$697 \rightarrow 523\ D_2$		

.

(701 – 799)

D_1	D_2	D_3
701 → 526	707 → 398	703 → 565 D_1
705 → 529 D_1	715 → 604	711 → 676
709 → 532	723 → 407 D_3	719 → 577 D_1
713 → 535 D_3	731 → 695 D_3	727 → 614
717 → 538	739 → 416	735 → 524
721 → 541 D_1	747 → 631 D_3	743 → 637 D_1
725 → 544	755 → 425 D_1	751 → 572
729 → 547 D_2	763 → 388	759 → 641 D_1
733 → 550	771 → 434	767 → 692
737 → 553 D_1	779 → 658	775 → 491 D_2
741 → 556	787 → 443 D_2	783 → 496
745 → 559 D_3	795 → 767 D_3	791 → 668
749 → 562		799 → 641 D_1
753 → 565 D_1		
757 → 568		
761 → 571 D_2		
765 → 574		
769 → 577 D_1		
773 → 580		
777 → 583 D_3		
781 → 586		
785 → 589 D_1		
789 → 592		
793 → 595 D_2		
797 → 598		

.

(801 – 899)

D_1

801 → 601 D_1
805 → 604
809 → 607 D_3
813 → 610
817 → 613 D_1
821 → 616
825 → 619 D_2
839 → 622
833 → 625 D_1
837 → 628
841 → 631 D_3
845 → 634
849 → 637 D_1
853 → 640
857 → 643 D_2
861 → 646
865 → 649 D_1
869 → 652
873 → 655 D_3
877 → 658
881 → 661 D_1
885 → 664
889 → 667 D_2
893 → 670
897 → 673 D_1

D_2

803 → 452
811 → 685 D_1
819 → 461 D_1
827 → 524
835 → 470
843 → 712
851 → 479 D_3
859 → 776
867 → 488
875 → 739 D_2
883 → 497 D_1
891 → 847 D_3
899 → 506

D_3

807 → 767 D_3
815 → 581 D_1
823 → 695 D_3
831 → 500
839 → 505 D_1
847 → 805 D_1
855 → 722
863 → 820
871 → 797 D_1
879 → 470
887 → 749 D_1
895 → 767 D_3

(**901 – 999**)

D_1	D_2	D_3
901 → 676	907 → 766	903 → 572
905 → 679 D_3	915 → 515 D_2	911 → 577 D_1
909 → 682	923 → 658	919 → 776
913 → 685 D_1	931 → 524	927 → 637 D_1
917 → 688	939 → 793 D_1	935 → 500
921 → 691 D_2	947 → 533 D_1	943 → 896
925 → 694	955 → 769 D_1	951 → 803 D_2
929 → 697 D_1	963 → 542	959 → 730
933 → 700	971 → 820	967 → 919 D_3
937 → 703 D_3	979 → 551 D_3	975 → 695 D_3
941 → 706	987 → 938	983 → 830
945 → 709 D_1	995 → 560	991 → 637 D_1
949 → 712		999 → 712
953 → 715 D_2		
957 → 718		
961 → 721 D_1		
965 → 724		
969 → 727 D_3		
973 → 730		
977 → 733 D_1		
981 → 736		
985 → 739 D_2		
989 → 742		
993 → 745 D_1		
997 → 748		

.

(1001 – 1099)

D_1	D_2	D_3
$1001 \rightarrow 751\ D_3$	$1003 \rightarrow 847\ D_3$	$1007 \rightarrow 767\ D_3$
$1005 \rightarrow 754$	$1011 \rightarrow 569\ D_1$	$1015 \rightarrow 857\ D_1$
$1009 \rightarrow 757\ D_1$	$1019 \rightarrow 545\ D_1$	$1023 \rightarrow 692$
$1013 \rightarrow 760$	$1027 \rightarrow 578$	$1031 \rightarrow 653\ D_1$
$1017 \rightarrow 763\ D_2$	$1035 \rightarrow 874$	$1039 \rightarrow 869\ D_1$
$1021 \rightarrow 766$	$1043 \rightarrow 587\ D_2$	$1047 \rightarrow 884$
$1025 \rightarrow 769\ D_1$	$1051 \rightarrow 901\ D_1$	$1055 \rightarrow 628$
$1029 \rightarrow 772$	$1059 \rightarrow 596$	$1063 \rightarrow 730$
$1033 \rightarrow 775\ D_3$	$1067 \rightarrow 901\ D_1$	$1071 \rightarrow 859\ D_2$
$1037 \rightarrow 778$	$1075 \rightarrow 605\ D_1$	$1079 \rightarrow 911\ D_3$
$1041 \rightarrow 781\ D_1$	$1083 \rightarrow 686$	$1087 \rightarrow 871\ D_3$
$1045 \rightarrow 784$	$1091 \rightarrow 614$	$1095 \rightarrow 659\ D_2$
$1049 \rightarrow 787\ D_2$	$1099 \rightarrow 928$	
$1053 \rightarrow 790$		
$1057 \rightarrow 793\ D_1$		
$1061 \rightarrow 796$		
$1065 \rightarrow 799\ D_3$		
$1069 \rightarrow 802$		
$1073 \rightarrow 805\ D_1$		
$1077 \rightarrow 808$		
$1081 \rightarrow 811\ D_2$		
$1085 \rightarrow 814$		
$1089 \rightarrow 817\ D_1$		
$1093 \rightarrow 820$		
$1097 \rightarrow 823\ D_3$		

.

(1101 – 1199)

D_1

1101 → 826
1105 → 829 D_1
1109 → 832
1113 → 835 D_2
1117 → 838
1121 → 841 D_1
1125 → 844
1129 → 847 D_3
1133 → 850
1137 → 853 D_1
1141 → 856
1145 → 859 D_2
1149 → 862
1153 → 865 D_1
1157 → 868
1161 → 871 D_3
1165 → 874
1169 → 877 D_1
1173 → 880
1177 → 883 D_2
1181 → 886
1185 → 889 D_1
1189 → 892
1193 → 895 D_3
1197 → 898

D_2

1107 → 623 D_3
1115 → 637 D_1
1123 → 632
1131 → 955 D_2
1139 → 886
1147 → 1090
1155 → 650
1163 → 982
1171 → 659 D_2
1179 → 1064
1187 → 668
1195 → 1009 D_1

D_3

1103 → 1048
1111 → 938
1119 → 1063 D_3
1127 → 1099 D_2
1135 → 910
1143 → 965 D_1
1151 → 692
1159 → 734
1167 → 739 D_2
1175 → 992
1183 → 1067 D_2
1191 → 955 D_2
1199 → 1139 D_2

.

83

(1201 – 1299)

D_1	D_2	D_3
$1201 \rightarrow 901\ D_1$	$1203 \rightarrow 677\ D_1$	$1207 \rightarrow 1019\ D_2$
$1205 \rightarrow 904$	$1211 \rightarrow 767\ D_3$	$1215 \rightarrow 974$
$1209 \rightarrow 907\ D_2$	$1219 \rightarrow 686$	$1223 \rightarrow 1162$
$1213 \rightarrow 910$	$1227 \rightarrow 1036$	$1231 \rightarrow 658$
$1217 \rightarrow 913\ D_1$	$1235 \rightarrow 695\ D_3$	$1239 \rightarrow 1046$
$1221 \rightarrow 916$	$1243 \rightarrow 1181\ D_1$	$1247 \rightarrow 1000$
$1225 \rightarrow 919\ D_3$	$1251 \rightarrow 704$	$1255 \rightarrow 637\ D_1$
$1229 \rightarrow 922$	$1259 \rightarrow 1063\ D_3$	$1263 \rightarrow 1185\ D_1$
$1233 \rightarrow 925\ D_1$	$1267 \rightarrow 713\ D_1$	$1271 \rightarrow 1073\ D_1$
$1237 \rightarrow 928$	$1275 \rightarrow 767\ D_3$	$1279 \rightarrow 730$
$1241 \rightarrow 931\ D_2$	$1283 \rightarrow 722$	$1287 \rightarrow 815\ D_3$
$1245 \rightarrow 934$	$1291 \rightarrow 1090$	$1295 \rightarrow 820$
$1249 \rightarrow 937\ D_1$	$1299 \rightarrow 731\ D_2$	
$1253 \rightarrow 940$		
$1257 \rightarrow 943\ D_3$		
$1261 \rightarrow 946$		
$1265 \rightarrow 949\ D_1$		
$1268 \rightarrow 952$		
$1273 \rightarrow 955\ D_2$		
$1277 \rightarrow 958$		
$1281 \rightarrow 961\ D_1$		
$1285 \rightarrow 964$		
$1289 \rightarrow 967\ D_3$		
$1293 \rightarrow 970$		
$1297 \rightarrow 973\ D_1$		

.

(1301 – 1399)

D_1

1301 → 976
1305 → 979 D_2
1309 → 982
1313 → 985 D_1
1317 → 988
1321 → 991 D_3
1325 → 994
1329 → 997 D_1
1333 → 1000
1337 → 1003 D_2
1341 → 1006
1345 → 1009 D_1
1349 → 1012
1353 → 1015 D_3
1357 → 1018
1361 → 1021 D_1
1365 → 1024
1369 → 1027 D_2
1373 → 1030
1377 → 1033 D_1
1381 → 1036
1385 → 1039 D_3
1389 → 1042
1393 → 1045 D_1
1397 → 1048

D_2

1307 → 797 D_1
1315 → 740
1323 → 1117 D_1
1331 → 749 D_1
1339 → 874
1347 → 758
1355 → 1144
1363 → 767 D_3
1371 → 977 D_1
1379 → 776
1387 → 1171 D_2
1395 → 785 D_1

D_3

1303 → 1100
1311 → 934
1319 → 1253 D_1
1327 → 1064
1335 → 1127 D_3
1343 → 767 D_3
1351 → 1219 D_2
1359 → 1291 D_2
1367 → 1154
1375 → 1306
1383 → 712
1391 → 1103 D_3
1399 → 1181 D_1

.

(1401 – 1499)

D_1	D_2	D_3
1401 → 1051 D_2	1403 → 1000	1407 → 1220
1405 → 1054	1411 → 794	1415 → 896
1409 → 1057 D_1	1419 → 1198	1423 → 901 D_1
1413 → 1060	1427 → 803 D_2	1431 → 1208
1417 → 1063 D_3	1435 → 767 D_3	1439 → 730
1421 → 1066	1443 → 812	1447 → 1031 D_3
1425 → 1069 D_1	1451 → 1225 D_1	1455 → 1382
1429 → 1072	1459 → 821 D_1	1463 → 1235 D_2
1433 → 1075 D_2	1467 → 929 D_2	1471 → 797 D_1
1437 → 1078	1475 → 830	1479 → 1405 D_1
1441 → 1081 D_1	1483 → 1144	1487 → 1274
1445 → 1084	1491 → 839 D_3	1495 → 1262
1449 → 1087 D_3	1499 → 1424	
1453 → 1090		
1457 → 1093 D_1		
1461 → 1096		
1465 → 1099 D_2		
1469 → 1102		
1473 → 1105 D_1		
1477 → 1108		
1481 → 1111 D_3		
1485 → 1114		
1489 → 1117 D_1		
1493 → 1120		
1497 → 1123 D_2		

.

(1501 – 1599)

D_1	D_2	D_3
1501 → 1126	1507 → 874	1503 → 827 D_2
1505 → 1129 D_1	1515 → 1279 D_3	1511 → 767 D_3
1509 → 1132	1523 → 857 D_1	1519 → 1370
1513 → 1135 D_3	1531 → 1091 D_2	1527 → 1289 D_1
1517 → 1138	1539 → 861 D_1	1535 → 1384
1521 → 1141 D_1	1547 → 1306	1543 → 977 D_1
1525 → 1144	1555 → 875 D_2	1551 → 982
1529 → 1147 D_2	1563 → 1253 D_1	1559 → 1316
1533 → 1150	1571 → 884	1567 → 1256
1537 → 1153 D_1	1579 → 1333 D_1	1575 → 1496
1541 → 1156	1587 → 893 D_1	1583 → 1256
1545 → 1159 D_3	1595 → 1010	1591 → 1343 D_3
1549 → 1162		1599 → 1139 D_2
1553 → 1165 D_1		
1557 → 1168		
1561 → 1171 D_2		
1565 → 1174		
1569 → 1177 D_1		
1573 → 1180		
1577 → 1183 D_3		
1581 → 1186		
1585 → 1189 D_1		
1589 → 1192		
1593 → 1195 D_2		
1597 → 1198		

.

(1601 – 1699)

D_1	D_2	D_3
1601 → 1201 D_1	1603 → 902	1607 → 1145 D_1
1605 → 1204	1611 → 1360	1615 → 1534
1609 → 1207 D_3	1619 → 911 D_3	1623 → 1370
1613 → 1210	1627 → 1468	1631 → 1549 D_1
1617 → 1213 D_1	1635 → 920	1639 → 1424
1621 → 1216	1643 → 1387 D_2	1647 → 880
1625 → 1219 D_2	1651 → 929 D_1	1655 → 1397 D_1
1629 → 1222	1659 → 1576	1663 → 1424
1633 → 1225 D_1	1667 → 938	1671 → 1058
1637 → 1228	1675 → 1414	1679 → 1063 D_3
1641 → 1231 D_3	1683 → 947 D_2	1687 → 1424
1645 → 1234	1691 → 1465 D_1	1695 → 1450
1649 → 1237 D_1	1699 → 956	
1653 → 1240		
1657 → 1243 D_2		
1661 → 1246		
1665 → 1249 D_1		
1669 → 1252		
1673 → 1255 D_3		
1677 → 1258		
1681 → 1261 D_1		
1685 → 1264		
1689 → 1267 D_2		
1693 → 1270		
1697 → 1273 D_1		

.

(1701 – 1799)

D_1

1701 → 1276
1705 → 1279 D_3
1709 → 1282
1713 → 1285 D_1
1717 → 1288
1721 → 1291 D_2
1725 → 1294
1729 → 1297 D_1
1733 → 1300
1737 → 1303 D_3
1741 → 1306
1745 → 1309 D_1
1749 → 1312
1753 → 1315 D_2
1757 → 1318
1761 → 1321 D_1
1765 → 1324
1769 → 1327 D_3
1773 → 1330
1777 → 1333 D_1
1781 → 1336
1785 → 1339 D_2
1789 → 1342
1793 → 1345 D_1
1797 → 1348

D_2

1707 → 1441 D_1
1715 → 965 D_1
1723 → 1091 D_2
1731 → 974
1739 → 1468
1747 → 983 D_3
1755 → 1667 D_2
1763 → 992
1771 → 1495 D_3
1779 → 1001 D_1
1787 → 955 D_2
1795 → 1010

D_3

1703 → 910
1711 → 1625 D_1
1719 → 1451 D_2
1727 → 1384
1735 → 1648
1743 → 1594
1751 → 1478
1759 → 1253 D_1
1767 → 1345 D_1
1775 → 1067 D_2
1783 → 1505 D_1
1791 → 1211 D_2
1799 → 1139 D_2

.

(1801 – 1899)

D_1	D_2	D_3
1801 → 1351 D_3	1803 → 1522	1807 → 1144
1805 → 1354	1811 → 1019 D_2	1815 → 1532
1809 → 1357 D_1	1819 → 1067 D_2	1823 → 974
1813 → 1360	1827 → 1028	1831 → 1739 D_2
1817 → 1363 D_2	1835 → 1549 D_1	1839 → 1310
1821 → 1366	1843 → 1037 D_1	1847 → 1559 D_3
1825 → 1369 D_1	1851 → 1172	1855 → 991 D_3
1829 → 1372	1859 → 1046	1863 → 1064
1833 → 1375 D_3	1867 → 1576	1871 → 1777 D_1
1837 → 1378	1875 → 1055 D_3	1879 → 1586
1841 → 1381 D_1	1883 → 1274	1887 → 1792
1845 → 1384	1891 → 1064	1895 → 1622
1849 → 1387 D_2	1899 → 1603 D_2	
1853 → 1390		
1857 → 1393 D_1		
1861 → 1396		
1865 → 1399 D_3		
1869 → 1402		
1873 → 1405 D_1		
1877 → 1408		
1881 → 1411 D_2		
1885 → 1414		
1889 → 1417 D_1		
1893 → 1420		
1897 → 1423 D_3		

.

(1901 – 1999)

D_1
1901 → 1426
1905 → 1429 D_1
1909 → 1432
1913 → 1435 D_2
1917 → 1438
1921 → 1441 D_1
1925 → 1444
1929 → 1447 D_3
1933 → 1450
1937 → 1453 D_1
1941 → 1456
1945 → 1459 D_2
1949 → 1462
1953 → 1465 D_1
1957 → 1468
1961 → 1471 D_3
1965 → 1474
1969 → 1477 D_1
1973 → 1480
1977 → 1483 D_2
1981 → 1486
1985 → 1489 D_1
1989 → 1492
1993 → 1495 D_3
1997 → 1498.

D_2
1907 → 1073 D_1
1915 → 1819 D_2
1923 → 1082
1931 → 1630
1939 → 1091 D_2
1947 → 1387 D_2
1955 → 1100
1963 → 1657 D_1
1971 → 1718
1979 → 1253 D_1
1987 → 1118
1995 → 1684

D_3
1903 → 1144
1911 → 1613 D_1
1919 → 1460
1927 → 1220
1935 → 1225 D_1
1943 → 1640
1951 → 1811 D_2
1959 → 1184
1967 → 1868
1975 → 1667 D_2
1983 → 1589 D_1
1991 → 1891 D_2
1999 → 1424

Summary

ㄲ.

Printed for Mnamon in the month of April 2015 at Andersen Spa Novara
Published by Mnamon in ebook version May 12, 2015
Published by Mnamon for on demand in Amazon.com on June 10th, 2015

www.ingramcontent.com/pod-product-compliance
Lightning Source LLC
Chambersburg PA
CBHW020158200326
41521CB00006B/418